Estimating and Cost Planning Using the New Rules of Measurement

Estimating and Cost Planning Using the New Rules of Measurement

Sean D.C. Ostrowski

WILEY Blackwell

Registered office: John Wiley & Sons, Ltd, The Atrium, Southern Gate, Chichester, West Sussex, PO19 8SQ, UK

Editorial offices: 9600 Garsington Road, Oxford, OX4 2DQ, UK
The Atrium, Southern Gate, Chichester, West Sussex, PO19 8SQ, UK
2121 State Avenue, Ames, Iowa 50014-8300, USA

For details of our global editorial offices, for customer services and for information about how to apply for permission to reuse the copyright material in this book please see our website at www.wiley.com/wiley-blackwell.

Library of Congress Cataloging-in-Publication Data
Ostrowski, Sean D. C.
Estimating and cost planning using the new rules of measurement / Sean D.C. Ostrowski.
pages cm
Includes bibliographical references and index.
ISBN 978-1-118-33265-8 (pbk. : alk. paper) 1. Building–Estimates. I. Title.
TH435.O845 2013
692′.5–dc23
2012037675

A catalogue record for this book is available from the British Library.

Wiley also publishes its books in a variety of electronic formats. Some content that appears in print may not be available in electronic books.

Cover image courtesy of Shutterstock
Cover design by Steve Thompson

Set in 10/12.5 pt Minion by Toppan Best-set Premedia Limited

1 2013

Contents

The book's companion website is at

www.wiley.com/go/ostrowski/estimating

You will find here freely downloadable support material

The author's website is at

http://ostrowskiquantities.com

Foreword

Good textbooks on explaining the principles and practice of building measurement are few and far between. Checking the catalogues of publishers will also reveal that only a few have longevity. A great many students have difficulty in developing the required skills so that they are able to properly and correctly apply them to new situations, a competence that is essential in the workplace. For some students the problem relates to a lack or poor understanding of construction technology. In the past I have often asked students whether they understand the principles of how buildings are constructed. Their replies are often encouraging but it is not long before I realise that their knowledge is pitifully inadequate. Without this knowledge no one will be able to adequately measure building works. For other students it is often due to a lack of study or an understanding of the rules of measurement. Others find it difficult to visualise construction in three dimensions and this inhibits their progress. These skills are very much in demand in practice and no amount of information technology has yet been able to replace them.

This is a new textbook that has been written with the above in mind by an author who I have worked with in delivering lectures on the New Rules of Measurement (NRM volume 1) on behalf of the RICS. I therefore know that he has a clear understanding of just what is required through many years of practice and teaching. He has recognised that the subject is difficult for students to understand and that insufficient time is allocated to its study in colleges and universities, as the demand for space in the curriculum exceeds the amount of time that is available. In order to help students overcome these difficulties he has therefore written this book interactively. This has never before been attempted to help facilitate an easier and better understanding of the way in which building works are measured. This will allow students to gain both a quicker and better grasp and understanding of the subject and the skills that are required to be able to apply the subject in practice whether working in a private office or with a contractor.

Furthermore the author has taken the trouble to try and understand *how* students acquire the knowledge, understanding and skills required for building measurement. He has given this much thought through a doctoral programme and has used his findings when writing this book. Once a student has properly acquired them, he or she will then have little difficulty in using them on other types of projects such as civil engineering or process plant engineering.

Professor Allan Ashworth
York, 2012

Preface

'Estimating and Cost Planning Using the New Rules of Measurement' is intended to provide some guidance on all the technical competencies concerned with estimating throughout the precontract period and this volume provides comprehensive and detailed examples of the work that is undertaken. The text and examples have been drawn from my professional and academic practice as a chartered quantity surveyor and lecturer.

The publication by the RICS of the suite of New Rules of Measurement (NRM) provides a prescriptive approach to the measurement of quantities throughout the construction process. The first of these, NRM 1, provides new methods of measurement for estimates and cost plans. As is often the case with innovations, some guidance can be useful and a commentary with examples and exercises on how to use NRM 1 is therefore appropriate. This textbook includes examples on how to measure estimates and cost plans in accordance with NRM 1.

The publication of the RICS Black Book guidance notes on acceleration and damages for delay to completion provides best practice to the quantity surveying profession and could, perhaps, indicate further developments in standard methods of measurement in an area very much in need of an accurate and consistent approach.

For both students and practitioners, the acquisition of technical competencies is by practice. A textbook can only provide an introduction. For this reason each chapter has a step-by-step worked example that can be followed and an opportunity to practise with an exercise on each topic. Cognitive development can be monitored by using the self-assessment marking sheets that are also provided.

A work like this will contain errors and they are entirely my own responsibility. I would be grateful for your assistance if you would be kind enough to please point out these errors and I will correct them at the first opportunity. I am also aware that some of the opinions in this volume will not be shared by all. I welcome your opinions which I will carefully consider.

Sean D.C. Ostrowski
Spring 2012

Acknowledgements

The author and the publisher would like to thank the following individuals and organisations for their kind permission to use the following materials: Professor Allan Ashworth for allowing me to use the material in Chapter 20 of *Cost Studies of Buildings* 5th edn, Prentice Hall Pearson to form Tables 8.8 and 8.9; G. Godwin/Builder Group/United business media for the use of W. Atton *Estimating Applied to Building*, 3rd edn, for the use of brickwork labour constants; CEM for the use of the drawings of the office building; the RICS and BCIS for allowing me to reproduce tables included in NRM 1, the code of measuring practice and elemental analyses and key performance indicators from Elemental Standard Form of Cost Analysis for the use of their table concerning floor areas; Standard Form of Cost Analysis; Taylor and Francis/Spon for the use of the brickwork pricing example in Chapter 9; and Turner and Townsend for the use of their diagrams concerning floor areas in Chapter 3.

Many people have read and commented on parts of the manuscript and they have my sincere thanks and gratitude. In particular, I wish to thank: my wife, Sally, for her assistance with proof reading; David Benge of Gleeds for our useful initial discussions on NRM 1; Keith Tweedy for his careful appraisal and discussions on the NRMs; Professor Allan Ashworth for his encouragement and tolerance; David Hockley, RIBA, and Matthew Boughton who despite busy workloads prepared many of the drawings and helped to convert them into the manuscript; Ian Pegg and Cosmos Kamasho of the BCIS; Professor David Jenkins and David Quarmby of Glamorgan University.

I also wish to thank the students of several institutions who diligently, and sometimes gleefully but always in good humour, pointed out everything that they could find that was wrong or inconsistent; also Madeleine Metcalfe, other members of staff and the technical assessors at Wiley Blackwell for their encouragement and comments for which I am profoundly grateful.

List of Tables and Diagrams

Glossary of Terms

APC	Assessment of Professional Competence
BCIS	Building Cost Information Services
BMS	building management system
BQs	Bills of Quantities
BREEAM	Building Research Establishment Environmental Assessment Method
BS	British Standard
BWIC	builders' work in connection
CAT	category
CCTV	closed circuit television
CITB	Construction Industry Training Board
CP	Cost Plan
DS	district surveyor
DDA	Disability Discrimination Act
DPC	damp proof course
F/M	foreman
GA	ground area
GEA	Gross External Area
GEFA	Gross External Floor Area
GIA	Gross Internal Area
GIFA	Gross Internal Floor Area
h/c	hardcore
IT	information technology
JCT	Joint Contracts Tribunal
Lab	labourer
M & E	mechanical and electrical
NEDO	National Economic Development Office
NIA	Net Internal Area
NIFA	Net Internal Floor Area
NRM	New Rules of Measurement
NUA	net useable area
O/H & P	overheads and profits
OGC	Office of Government Commerce
PAFI	price adjustment formulae indices
PFI	price fluctuation index
PPP	public–private partnership
RIBA	Royal Institute of British Architects
RICS	Royal Institution of Chartered Surveyors
SMM	standard method of measurement
ssw	sawn softwood
T & C	test and commission
TPI	Tender Price Index
VAT	Value Added Tax

Glossary of Terms

[illegible]

BREEAM Building Research Establishment Environmental Assessment Method

[illegible]

VAT Value Added Tax

1 Introduction

1.1 INTRODUCTION

Buildings have become complex machines for specialist functions and the materials and construction processes for the construction of these buildings have become largely the erection of prefabricated materials and components. The need for the construction process to be integrated by the collaboration of all the parties involved is now the major management requirement. The skill levels of craftsmen has been superseded by the management resources necessary to ensure a coordinated approach across the interfaces between the client, designers, cost management, contractors and subcontractors.

The cost management part of this process should also reflect this development with a comprehensive framework which is consistent from the beginning to the end of the construction cycle and is transparent enough to be understood at any time by all the stakeholders. This need for collaboration has not been a priority until recently. Seeley (1976) described approximate estimating as: '. . . *techniques which attempted to give a forecast of the probable tender figure, although the basis of the computation often left much to be desired*'. (p. 1). He went on to explain that the method of approach was often dictated by the nature of the development and the nature of the promoter. There were different kinds of estimates for different kinds of projects. Every professional practice and contractor had its own rules. This helps to explain why there have been no rules on how to

Estimating and Cost Planning Using the New Rules of Measurement, First Edition. Sean D.C. Ostrowski.

prepare estimates. This has led to complaints from client organisations that they receive a different estimate at different stages and from different consultants. Many client bodies have become active partners in the construction process and these knowledgeable employers administer the contracts from beginning to end. They have become aware of the inconsistencies in financial reporting throughout the construction cycle.

The Royal Institution of Chartered Surveyors (RICS) now provides the following documents that will help to provide comprehensive, accurate and consistent financial reporting:

- *The RICS Code of Measuring Practice*, 6th edition, 2007
- *The RICS New Rules of Measurement NRM 1: Order of Cost Estimating and Cost Planning for Capital Works*, 2nd edition, April 2012
- *The RICS New Rules of Measurement NRM 2: Detailed Measurement for Building Works*, April 2012
- *The RICS New Rules of Measurement NRM 3: Maintenance and Operations Cost Estimating, Planning and Procurement*, to be published in 2013
- *BCIS Elemental Standard Form of Cost Analysis*, 4th edition, 2012

The New Rules of Measurement (NRM) provide an accurate and consistent approach through the full life cycle of the building at each stage of development: the estimate; the cost plans; the work packages and bills of quantities and finally the whole life costing maintenance programmes. The intention is to provide a transparent audit trail of the measurement and the pricing that is available from the beginning to the end of a building's life cycle. There is an expectation that the NRM will have a wide appeal with an opportunity for countries around the world to adopt a common set of rules for the measurement of building works.

Contents

This textbook examines and explains how to use the *RICS New Rules of Measurement, NRM 1: Order of Cost Estimating and Cost Planning for Capital Works*, 2nd edition April 2012 and includes an introduction to all the major components of estimating necessary to produce an anticipated cost for a contract. It is intended for use by students and practitioners. The contents are:

1. Introduction
2. A practical introduction to measurement
3. The Code of Measuring Practice
4. How to use the NRM
5. NRM estimates
6. NRM cost plans
7. Information
8. Preliminaries, risk, overheads and profit
9. Unit rates
10. Cost analysis

Also included is:

- A detailed worked example of the practical application in most chapters
- A comprehensive exercise for practice at the end of most chapters

- A detailed answer with the calculations
- A self-assessment marking sheet to provide an indication of the standard achieved in technical and managerial competence and cognitive development

RICS competence levels

The RICS Assessment of Professional Competence (APC) comprises the demonstrable acquisition of a series of competencies after a period of time in the profession. This includes the provision of a diary showing a structured training programme and a final assessment interview. Two of the competencies are: 'design economics and cost planning' and 'quantification and costing of construction works'. The first concerns estimating and cost planning and the second concerns measurement. They are core technical competencies that are mandatory for the successful completion of the APC. Each competence has three levels. Level 1 is knowledge about the subject. This is the provision of propositional knowledge about the subject. Level 2 is being able to apply that knowledge. This is the provision of procedural knowledge, actually being able to undertake the competence to a level of skill that is both comprehensive and accurate. Level 3 is being able to discriminate the quality of the work and advise the client.This will only be available after some time in practice, which is part of the RICS APC programme. This textbook provides worked examples can be followed and replicated. Practical exercises at the end of the chapters provide practice at Level 2. The marking scheme provides some level of discrimination on the quality of the work that has been practised. How to use these NRMs is therefore a necessary step on the route to Level 3.

Practical examples and self-assessment exercises

The pedogogy of quantity surveying is primarily concerned with the nature of the knowledge that is being learnt. Technical competencies, like estimating, measurement and contract administration are procedural knowledge. Practice is the most useful method of teaching procedural knowledge (Gagne, 2002) and teaching, in the form of telling or demonstrating to the student how to do it, is not effective (Wood, 2001). The acquisition of a technical competence like measurement is learnt by practice. The use of textbooks, lectures or demonstrations only provides an introduction to how to measure. The most effective way of learning how to measure is to practise and to receive prompt answers to questions as they arise. However, before the practice can take place it is helpful to examine previous examples to see the process that is required and how to set out the work. An examination of the examples at the end of each chapter will provide the information necessary to carry out the self-assessment exercises.

The practical examples and self-assessment exercises are set out on the traditional rulings used for estimating paper. Although much work can now be done on spreadsheets the need to understand the construction technology, the use of side casts and the conversion calculations to enable the measurement to be compliant with the NRM are all easier to understand if set out on estimating paper. Several of the examples and exercises refer to a set of common drawings for the London Road office project. These drawings will be used in Chapters 4–6. This enables the vocabulary, technology and dimensions to be acquired progressively. When the competence levels have reached Level 2, an ability to carry out the work comprehensively and accurately without supervision, then the alternative technologies can be introduced. Software measurement packages require a significant amount of practice before they can be used effectively. Proficiency in using the software is best acquired after expertise in measurement has been attained.

Companion websites

Many students, particularly at the outset, find the subject difficult. The printed word in the form of a textbook has a limited usefulness in providing the appropriate teaching for these technical competencies. The most effective method of acquiring expertise in these disciplines is by practice and the contemporaneous answers to questions as they arise. To provide further assistance there are dedicated websites at http://ostrowskiquantities.com and at Wiley Blackwell (http://www.wiley.com/go/ostrowski/estimating). It is hoped that the provision of this will go some way towards expanding the opportunities for practice in a more useful way than using the printed word alone.

The RICS website (www.rics.org) includes the NRMs free of charge for members and they also available on the RICS subscription information service (ISurv) (www.rics.org/uk/knowledge/more-services/professional-services/isurv/). Most practices, contractors and universities are subscribers. This means that a screen-based version is available to most individuals free of charge and hard copies can be obtained for the cost of the printing.

1.2 STANDARD METHODS OF MEASUREMENT

There are currently several different methods of measurement published by the RICS as follows:

- The RICS Code of Measuring Practice. This sets out how to measure floor areas.
- NRM 1, The RICS New Rules of Measurement for Estimating and Cost Planning. This provides a method of measurement for quantities on an elemental basis for estimates and a different method of measurement for cost plans.
- NRM 2, The RICS New Rules of Measurement for Building Works. This measures quantities on a trade basis.
- NRM 3, The RICS New Rules of Measurement for Whole Life Costs Concerning Maintenance, Renewal and Inspection. This measures quantities on an elemental basis as NRM 1.

The introduction of the full suite of NRMs means that there are accurate forms of measurement that will need to be prepared to reflect the design and specification at the end of significant stages of the design process. They should also provide accurate prices from the information made available.

Elemental measurement

The first task is to measure the work using an elemental method of measurement such as NRM 1 (Order of Cost Estimating and Cost Planning for Capital Building Works). Each element of the building is measured, eg a reinforced concrete roof will comprise several trades, *viz.* concrete, formwork, reinforcement, screed, asphalt, metalwork, balustrading etc. Within this elemental method there is a different method of measurement for the estimate and the cost plan. The estimate uses preparatory design information and mainly uses the superficial floor areas as the basis for measurement. The cost plans uses a progressively more developed design and measure quantities using units that are cubic superficial, enumerated and itemised for a larger range of elements.

Trade measurement

The second task is to measure the work using the trade method of measurement in NRM 2 (Detailed Measurement for Building Works). Each trade is measured separately wherever it occurs in the building, using a technically complete design. This enables the efficient collection of trade works into separate bills of quantities and ease of pricing by the contractor.

Compatibility

The goal of a strong, seamlessly linked, cost control pathway has commenced with the publication of NRM 1 and 2 which provide sets of rules that are accurate and consistent. It can also be seen that NRM 1 and 2 provide more than one set of rules which are alternative and overlapping methods to measure quantities. The progressive measurement and pricing stages, for estimates, cost plans and trades provides a structured cost management framework that is more detailed and accurate at each stage. However, standardisation, accuracy and consistently will be lost if the prices in the estimates and cost plans which have been prepared on an elemental basis using NRM 1 are not compatible with the tendered prices which have been prepared on a trade basis using NRM 2. The audit trail will be disconnected and the transparency that the client requires may not be possible. The measurement and pricing in the cost plans cannot be compared with the measurement and pricing in the trade bills of quantities and the client may consider that going out to tender on the elemental measurements included in the cost plans will provide adequate early stage prices. However, comprehensive financial security for the client can be made available with the measurement of trade bills of quantities using NRM 2 to be priced by the contractor to provide a fully quantified schedule of rates. In this way the perception that the bills of quantities are a barrier to collaboration amongst the stakeholders is removed.

It is possible to combine the advantages of both NRM 1 and NRM 2 into a single structured set of rules. The advantage of elemental measurement is the relevance of the costs to a particular part of the building. The advantage of trade measurement is that this is the basis of the pricing for the contractors and subcontractors who construct the work. Weights and volumes will remain the basis of pricing substantial parts of all construction work. The appropriate parts of the trade bills of quantities can be allocated to the appropriate elements.

1.3 PRICING

Accurate prices

The perennial problem with estimates is that they are not accurate enough. Despite the vast amount of information that is available this problem remains. Some research has been carried out by an eminent cost engineer in the USA. Hackney (1992) refers to a Rand study in Chapter 53 concerning the quantified effects of management decisions. He states that: '*Estimates prepared by groups with a vested interest in having the projects approved were found to be associated with added cost growth. Cost growth for their project averaged 49%, compared with 22% for estimates prepared by estimators independent of the group sponsoring the project . . .*' (p. 456).

This confirms our impression that all estimates are problematical. He goes on to state the reasons why: '*The primary reason for this difference appears to be that the average estimate prepared by project champions were less well defined than the average estimate prepared by independent estimators . . . . The difference may also reflect the tendency of sponsor groups to provide their estimators with optimistic assumptions.*' (p. 456–7).

Complex buildings require complex estimates and this is often in advance of sufficient information being made available. The ability to predict an accurate price is restricted to the quality and quantity of the information available and the expertise and experience of the surveyor. As a consequence the anticipated costs should be expressed as a range rather than a single target figure.

2 A Practical Introduction to Measurement

2.1 A PRACTICAL INTRODUCTION TO MEASUREMENT

Measurement protocols

The measurement of these quantities requires several sets of documents to be available at the same time, the drawings, specifications, the NRM and the measurement system (software or manual). They may all be screen-based and access to each document requires excessive transfer between screen images. To reduce this it is recommended that the drawings and NRM are used in a hard copy format.

The following example (see Table 2.1) sets out a typical estimating sheet. Each item is numbered and explained as follows:

1. 'Rulings' is the collective name for the various kinds of layouts used by the quantity surveyor and refers to the type of vertical lines used on the paper. This example is estimating paper.
2. Every page has 'headers' to identify the project and 'footers' for page numbers.
3. The dimension column sets out the dimensions in metres to two decimal points.
4. The 'timesing' column provides any multiples that are necessary.
5. The 'squaring' column gives the product and sum of each of the dimensions.

Estimating and Cost Planning Using the New Rules of Measurement, First Edition. Sean D.C. Ostrowski.

Table 2.1 Measurement protocols.

4	3	5	2 / 1 / 6	8	9	10	
			Basement Cost Plan				
	12 48.00 12.00 3.50	2,016.00	Basement excavation [1.1.4.1 & 7 Disposal of excavated material [1.1.4.2	2,016m^3 2,016m^3	10.00	20,160	00
2/ 2/	13 48.00 3.50 12.00 3.50	336.00 84.00 420.00	11 10/4800 48000 2/6000 12000 Earthwork support [1.1.4.4	420m^2			
15 3.5/	14 36.00	288.00	Designed joints [2.1.4.5	288 m			
4/	16 1	4	Pile tests [1.1.2.10	4 nr			
	17 Item	18	Dewatering [0.4.1.1	Item			
			23				

6. This column provides the description of the item using the appropriate elements and levels set out in the NRM.
7. This is the NRM reference.
8. This is the final quantity expressed as an integer, a whole number with the unit of measure.
9. This column is the price or 'rate' for each item.
10. The final column is the price expressed as the product of the quantity multiplied by the rate.
11. Calculations that may be necessary to provide a dimension are 'side-casts' (also called 'waste calculations'). They are dimensions shown on the drawings which are always shown in millimetres.
12. Cubic dimensions: length × width × depth.
13. Superficial dimensions: width first multiplied by height.
14. Linear dimensions.
15. Dotting on allows further multiples of dimensions without repeating the dimension.
16. Enumerated dimensions.
17. Items that are not measured.
18. Brackets to contain sets of dimensions and/or descriptions.

Scales

Drawings can come from a variety of sources, free-form sketches, hand-drawn plans, elevation and sections on paper and several software packages that make two- and three-dimensional drawings. Typical scales are:

- 1:1250 small-scale drawings to show large areas, eg landscaping
- 1:100 design drawings for layouts and elevations, eg plans
- 1:50 production drawings to show the construction, eg sections
- 1:5 large-scale drawings to show details, eg interfaces

At the estimating stage scales may not be available. At the cost planning stage the scales and some figured dimensions are shown on the drawings. The information requirements set out in NRM 1 do not specifically detail scale requirements. However, NRM 2 states at clause 2.14.4.1 *'Drawings shall be to a suitable scale'.* The extensive electronic transmission between terminals and software packages and printing on different sizes of paper will often make these scales unusable. This reduction and enlargement of drawings is common between computer terminals and printers and care needs to be taken to ensure that the scales shown on the drawings are not used to extrapolate dimensions that are not shown as figured dimensions. The scale of the drawings is one of the first matters that should be checked.

For our working examples and exercises throughout this book the drawings often distort the dimensions to such an extent that they cannot be scaled. Figured dimensions on the drawings are the only dimensions that should be used. The improvement that technology has provided in the rapid electronic exchange of drawings has been offset by the problems of changes in scale with enlargements and reductions of the printed drawings. Dimensions that are scaled or extrapolated should be confirmed before being used in published documents. These dimensions can be included on a query sheet similar to the example provided in Chapter 3 (Table 3.8) and sent to the architect for confirmation. All dimensions should be figured dimensions provided by the designers.

Accuracy

One of the major anxieties that students encounter is how accurate should these measurements be? The desire for accuracy to the nearest millimetre is often the first step of the diligent student. However, this is not necessary. Guidelines set out in NRM 2 can be used for the preparation of estimates and cost plans. NRM 2, p. 46, item 3.3.2(d), states that dimensions used in calculating quantities shall be taken to the nearest 10 mm. Item 3.3.2(e) states that quantities shall be to the nearest whole number. This can be illustrated by measurement in cubic metres.

A dimension of $3460 \times 3460 \times 3460 = 41.42$ m^3. This is 41 m^3 to the nearest whole number
A dimension of $3500 \times 3500 \times 3500 = 42.88$ m^3. This is 43 m^3 to the nearest whole number
A dimension of $3525 \times 3525 \times 3525 = 43.80$ m^3. This is 44 m^3 to the nearest whole number

This demonstrates that a range of 65 mm from 3460 to 3525 only increases the cubic quantity by 3 m^3.

Symbols

The NRM (Section 1.6 Symbols, abbreviations and definitions) gives a specific set of symbols which are to be used. Take as an example the symbol 'No.' to indicate an enumerated item like 5 No. manholes. This may be your normal method for describing enumerated items. However, the NRM says that this should be shown as 5 nr manholes. There is no capital or decimal point. The point as a decimal marker (00.11) and the comma as a thousand separator (10,000.00) are almost universally used in construction. However, the manufactured products used in construction often use a different convention. To abandon the use of the normal construction conventions would render the quantities in any bills of quantities confusing and misleading. The Code of Measuring Practice provides a timely warning. '*The British Standard BS 8888: 2006 Technical Product Specification (for defining, specifying, and graphically representing products) recommends the inclusion of a comma rather than a point as a decimal marker, and a space instead of a comma as a thousand separator. While the convention has not been adopted in this Code, users should take care to ensure that this does not conflict with client requirements*'. (Code of Measuring Practice, 6th edition 2007, p. 5).

Query sheet/to-take lists/marked-up drawings

The expertise that quantity surveyors have acquired enables them to examine the drawings and highlight areas that need clarification, additional information or possible corrections. Where there are queries on the drawings they should be scheduled on a query sheet. By providing the query sheets to the design team they are assisting the design process to the benefit of all concerned.

A 'to-take' list to act as an *aide-mémoire* to ensure that nothing is forgotten will be of assistance in ensuring that each item is assigned to the correct element. The intent of the elemental format of NRM 1 is that all the work in that element is included. This is helpful because it leaves no doubt that everything is to be included. For example, the roof may include the concrete slab, asphalt roofing, metalwork support structures and brickwork plant rooms. However, the plant in the plant room housed on the roof will be included as part of one of the services elements.

'Marked-up' drawings to demonstrate that all the work has been measured are a simple and effective way to ensure that nothing has been left out.

Revisions

The electronic transmission of revised drawings without a revision being timed or dated and partial alterations to parts of existing drawings means that the particular drawing or detail that is being used should be carefully recorded on the drawing schedule, prepared by the quantity surveyor, and noted on the dimensions.

Information

A further innovation is that the NRM now include extensive and detailed schedules of information for the provision of drawings and specifications. This should help to improve the information that is necessary for accurate estimates and cost plans and that it will be available in sufficient time.

2.2 MEASUREMENT PROCEDURE

Technical competence

Measurement is a technical competence using procedural knowledge. This is the knowledge of how to do things, how to carry out functional activities. Examples include riding a bicycle, playing the piano and bricklaying. The most important requirement of this work is to practise the procedure to enable improvement and expertise. Increased competence with the procedure comes with practice. The practice includes a procedure or process similar to a recipe for making a cake or addressing a golf ball. The elements are precisely defined in terms of quality and quantity and they occur in a predetermined sequence. It does not require any innovative thinking and it is not necessary to know the solution before the process commences. A successful outcome is simply a matter of following the process. The more it is practised the higher the level of competence that is acquired.

Procedure

Problems are often manifest in the form of the question 'What do I do next?' A procedure is set out below to address this question.

- The front page is the title page. It describes the project, the NRM section, the drawings, the specification, the author and the date.
- 'Top and tail' each page. The reference should be at the top and pagination at the bottom.

- Prepare a 'to-take' list from the NRM of all the items that need to be measured for that element.
- Look at the drawings to ascertain that all the required information is available.
- Prepare a query list to provide outstanding information.
- Calculate the 'side-casts' that are likely to be used: eg centre lines, floor-to-ceiling heights, room dimensions.
- Measure the work progressively using the NRM as a guide. This can be one of the following options:
  - all the items in the sequence of construction
  - all the items in each trade
  - all the items on each drawing
- Clear each item from the 'to-take' list.
- Mark up the drawings so that all the items are seen to have been measured.
- Prepare a list of outstanding items to be measured when clarification is provided.

Compound items

NRM 1 provides relatively simple methods of measurement. The intention is that several items are incorporated into the pricing of the work. However, these items still have to be carried out and the estimate or cost plan still has to include the cost of the work in the price. This requires the build up of a compound item which includes more than one unit of measurement, each one of which has a separate price. Two examples, the first for floor finishes and a second for a basement slab will illustrate the process.

An estimate for a suspended floor finish (NRM 1, p. 30, item 2.3.2) (see Table 2.2) includes the raised access floor, the floor finish and the skirting. To provide a comprehensive rate it is necessary to build up a compound unit rate comprising each of the items. Compound items also occur in cost plans but in this case each item is measured. Details of the measurement and pricing are not included at this stage.

Table 2.2 Compound items for estimates: Floor finishes.

(NRM 1, p. 30, item 2.3.2)			
Suspended floor		2,700 m^2 @ £30/m^3	81,000.00
Carpet tiles		2,700 m^2 @ £15/m^2	40,500.00
Skirting @ £5/m	£30 + £15 = £45/m^2 + say 5% for skirting. Say £2/m^2	2,700 m^2 @ £2/m^2	5,400.00
NRM 2.3.2	**Compound item for floor finishes**	2,700 m^2 @ £47/m^2	**£126,900.00**

The cost plan requires separate items to be measured and priced to build up the cost of this sub-element, as shown in Table 2.3.

Table 2.3 Measured items for cost plans: Floor finishes.

(NRM 1, p. 144–5, items 3.2.2.1, 3.2.1.1 and 3.2.1.3)			
3.2.2.1	Suspended floor	2,700 m^2 @ £30/m^3	81,000.00
3.2.1.1	Carpet tiles	2,700 m^2 @ £15/m^2	40,500.00
3.2.1.3	Skirting	700 m @ £5/m^2	3,500.00
			£125,000.00

An estimate for a reinforced concrete basement slab is measured in square metres (NRM 1, p. 27, item 1.1.2). The reinforcement, waterproof tanking and drainage below the slab are included. However, this work still has to be carried out and the estimator still has to include all the work in the measurement and pricing by incorporating all the items into the concrete slab item. This is carried out by building up a compound item. This measures separately each trade that is necessary to complete the work and applies a rate to each item. The sum of all the items is the total cost of the works necessary to carry out the work. This cost is then divided by the unit of measurement as set out by the NRM. For instance, a basement slab is measured in square metres as NRM item 1.1.2. However, concrete is measured and priced in cubic metres and reinforcement is measured and priced by weight. It is necessary to build up a compound item as shown in Table 2.4.

Table 2.4 Compound items for estimates: Basement slab.

NRM 1, p. 27 item 1.1.2			
Reinforced concrete slab		100 m^3 @ £100/m^3	10,000.00
Reinforcement @150 kg/m^3		15 t @ £2,000/m^2	30,000.00
Asphalt waterproofing/waterbar		400 m^2 @ £50/m^2	20,000.00
Cast iron drainage below slab		400 m^2 @ £10/m^2	4,000.00
			£64,000.00
		Area of basement slab is 400 m^2	÷400 m^2 = £160/m^2
NRM 1.1.2	**Compound item for basement slab**	400 m^2 @ £160/m^2	**£64,000.00**

The cost plan requires separate items to measured and priced to build up the cost for this sub-element as shown in Table 2.5.

Table 2.5 Compound items for cost plan: Basement slab.

NRM 1, p. 95–6, items 1.1.3.1, 1.1.3.3, 1.1.3.6 and 1.1.3.7			
1.1.3.1	Reinforced concrete slab	100 m² @£100/m³	10,000.00
	Reinforcement @150 kg/m³	15 t @ £2,000/t	30,000.00
	Asphalt waterproofing	400 m² @ £50/m²	20,000.00
£60,000 ÷ 400 m² = £150/m²		NRM 1.1.3.1 400 m² @ £150/m²	60,000.00
1.1.3.3	Extra over for lift pits	4 nr @ £500	2,000.00
1.1.3.6	Designed joints	100 m @ £15/m	1,500.00
1.1.3.7	Cast iron pipework below slab	50 m @ £100/m	5,000.00
1.1.3.9	Manholes	5 nr @ £1,000	5,000.00
1.1.3.10/11	Test and commission		Included
			£73,500.00

These examples indicate that the simplification of the method of measurement does not reduce the amount of measurement and pricing. The task of the estimator is to adjust the measurement and pricing to provide a comprehensive price. Simplification has come at a price which may not be cost-effective.

3 Code of Measuring Practice

3.1 INTRODUCTION

The purpose of the Code is to provide accurate and consistent measurement of floor areas with the overriding intention to provide clarity and certainty. There are three different ways of measuring floor areas. The use of a particular method of measurement is dependent on the type of building and the application for which it is intended. Some examples follow:

- gross external floor areas (GEFAs)
  eg for domestic buildings, for use with planning applications
- gross internal floor areas (GIFAs)
  eg for industrial premises, also applied to estimates and cost plans

Estimating and Cost Planning Using the New Rules of Measurement, First Edition. Sean D.C. Ostrowski.

- net useable areas (NUAs)
  eg for retail premises, also applicable to rent valuations

The measurement of these areas concerns accuracy, scales, the stage of the development, changes, complexity and transparency. Floor areas have a high level of accuracy with tolerances of below 1%. The scales on the drawings are problematic and have to be confirmed in advance of publication. The development of the floor areas through the construction process can become complex which means that the best advice that can be given is to maintain sufficient transparency for the client to be able to understand the way in which the areas have been measured and so that any changes that have been made can be followed. Good practice includes the definition of what is included and excluded and this requires the careful measurement of interfaces between elements in accordance with the Code. Various examples are provided to illustrate this.

An example of the practical application of the Code to measuring the floor area for a commercial office building is provided to allow the procedure to be understood. Using the rules in the Code the GIFA has been measured on estimating paper. A narrative has been included in italics to provide explanations. A self-assessment exercise is provided to allow practice of the technical competence of measuring areas.

3.2 THE PURPOSE OF THE CODE

A significant part of the work of a surveyor is concerned with measuring. The important factor concerning measurement is that it should provide a common standard that everyone will refer to and use. For example the people involved in constructing a building will want to know the size of the building in different ways. The client will want to know how much space will be available; the architect will want to know how to arrange the space; the structural engineer will want to know the weight of the components; the services engineer will want to know the volume of the building; the surveyor will want to know the floor areas. The purpose of the Code of Measuring Practice is to provide a particular method of measurement of floor areas for a particular application. This will enable everyone to be able to say that the floor areas for, say, an industrial unit will always be measured using the GIFA.

Accurate and consistent

The Code of Measuring Practice provides one of the basic elements in the work of surveyors. It provides the definitions of different kinds of floor areas and tells us how to measure these floor areas. This provides both the client and the professionals with a common and trustworthy schedule of floor areas which all parties can rely on from the beginning of the construction process right through to the end of the building's life cycle. The Code of Measuring Practice was first published by the RICS in 1979 and the 6th edition was issued in 2007. The purpose of the Code is set out in its Introduction (p. 1) '*. . . to provide succinct, precise definitions to permit accurate measurement of buildings and land, the calculation of the sizes (areas and volumes) and the description or specification of land and buildings on a common and consistent basis.*'

This purpose highlights that any form of measurement relies on accuracy and consistency. The accuracy is provided by providing a series of methods of how to measure the floor areas of buildings. The consistency comes from providing the definitions of different types of buildings. Each building type has a description or specification and each has a method of measurement that is applicable to it.

Intent

The intent of the Code is to provide a clear and unambiguous measurement. There are many different types of buildings and they do not always fit neatly into a particular definition. This means that the use of mixed layouts, calculations and formulae may be necessary and the application of these rules becomes quite technical and complex. Like any complex set of calculations they tend to develop ambiguities and anomalies which have come about for a variety of reasons. They are familiar to the professional measurers who may use them frequently, but they can be the cause of confusion and can be misleading to the client who may not be familiar with these complexities.

This brings us to another purpose of the Code which should be just as important. The intent of the Code is to provide floor schedules that do not confuse or mislead the recipients. The use of complex layouts, calculations and formulae may be necessary but the intent of the measurement should be that the schedules of areas provide clarity and certainty. The Code is quite clear on this matter and provides guidance for the appropriate duty of care as follows:

> *'In its response to a previous draft consultation paper, the Institute of Trading Standards Officers pointed out the line likely to be adopted by the courts. This will be that it does not matter what the professionals may think and understand, it is what the average person thinks and believes that is important in deciding whether the statements are misleading or not.'* (Code of Measuring Practice, 6th edition, 2007, p. 3).

Definitions

The Code provides core definitions of three different ways of measuring floor areas. They all relate to the extent to which the external and internal walls are included in the measurement of the areas. An easy way to remember the sequence is to consider each method of measurement as a series of concentric rings starting from the largest on the outside to the smallest on the inside. The gross external floor area (GEFA) measures the floor area from the outside face of the external walls and across all walls and partitions. The gross internal floor area (GIFA) measures the floor area from the inside face of the external walls and across all walls and partitions. The net internal floor area (NIFA) measures the useable floor area from the inside face of the external walls and between the walls and partitions. The definitions from the Code (pp. 8, 12 and 16) are more succinct as follows:

Gross External Floor Area (GEFA)

Gross External Area (GEA) is the area of a building measured externally at each floor level.

Gross Internal Floor Area (GIFA)

Gross Internal Area (GIA) is the area of a building measured to the internal face of the perimeter walls at each floor level.

Net Internal Floor Area (NIFA)

Net Internal Area (NIA) is the useable area within a building measured to the internal face of the perimeter walls at each floor level.

Specifications

These different methods of measurement are for buildings that are used for particular applications. In the Code, they are called descriptions or specifications. A few examples will illustrate the meaning at this stage. GEFAs are used in planning applications because they show the full extent of the building. GIFAs are used in estimates for the cost of buildings because they show the extent of the floor areas to be constructed. NIFAs are used in commercial valuations because they show the NUA of floor space available. The major types of applications are summarised in Table 3.1.

Table 3.1 Types of measurement and their application for different uses.

GROSS EXTERNAL AREA (GEA)	GROSS INTERNAL AREA (GIA)	NET INTERNAL AREA (NIA)
APPLICATIONS	APPLICATIONS	APPLICATIONS
Planning	Estimates	Commercial valuations
Council Tax	Industrial valuations	Retail valuations
Insurance	New home valuations	

Gross external area

Planning applications; insurance for a house and the Council Tax on domestic premises are based on the GEA.

Gross internal area

Estimates and cost plans for construction work use GIA. This application will be used throughout the remainder of the chapters for examples. The valuation of industrial premises and new homes also uses the GIA.

Net useable area

Valuations for commercial and retail properties use the NUA where the actual floor space available for commercial use is the significant information that the client requires.

3.3 MEASUREMENT

Having established the basic parameters concerning definitions, the appropriate methods of measurement and their particular applications, there are further items that should be taken into consideration.

Accuracy in floor plans

How accurate should these measurements be? Let us take for example an educational building of 13,000 m^2 and examine tolerance levels of 1% and 5% for accuracy to see if this would provide us with the appropriate levels of accuracy (Table 3.2).

Table 3.2 Tolerances of accuracy.

Total cost	13,000 m^2 @ £2,000/m^2 = £26,000,000	
	An area of 1% is 130 m^2	An area of 5% is 650 m^2
Construction cost	130 m^2 @ £2,000/m^2 = £260,000	650 m^2 @ £2,000/m^2 = £1,300,000
Rental income	130 m^2 @ £350/m^2 = £45,500	650 m^2 @ £350/m^2 = £227,500

A 1% tolerance level for the floor area may seem to be quite an accurate measurement. However, it is 130 m^2 and this is an area that is a substantial amount of space with a considerable construction cost of £260,000 and a rental income of £45,500. This 1% level of tolerance is not satisfactory for our purposes. By comparison the 5% tolerance level is quite unacceptable.

Table 3.3 is an example of a floor schedule and we can examine it to illustrate the information it provides and the extent of the accuracy and the appropriate level of accuracy. In this schedule of areas the method of measurement is the GIA. The area has been measured from the inside face of the external walls and across all partitions to obtain a total area of 8,225 m^2.

To ensure that this is accurate a check is carried out. First, the area of each room is measured (the NIA). This will not provide the full GIA because the area of the internal walls will not have been measured. The area of the internal partitions is then measured which is the length of the walls multiplied by the nominal width. The GIA is the internal area of the rooms, the NIA, plus the area of the partitions. The area of the internal walls is the balance between the NIA of the rooms and the full GIA. In the table the GIA is 8,225 m^2 and the internal walls are 494 m^2. The NIA is 8,225 − 494 = 7,731 m^2. The area of the frames and partitions is 494 m^2 which is 6% of the total and has been used as a balance to obtain the total area of 8,225 m^2.

In this example individual stair cores have been measured. The smallest is 16 m^2 which is 0.25% of the total. Print rooms have also been measured. The smallest is 3 m^2. It is unlikely that these dimensions will be available on the drawing at this stage and they may be derived from a schedule of accommodation provided by the client or architect or from scaling. The print room will need to have dimensions similar to 2.00 × 1.50 to get to 3 m^2. Using dimensions that are this small at this stage is problematical. The use of a balance allows the room sizes to be adjusted at a later stage without compromising the GIA.

This level of accuracy provides the client with the transparency and information that is needed to refine the schedule of accommodation and allows the quantity surveyor to demonstrate the accuracy of the figures (see Table 3.3).

Table 3.3 Schedule of Areas.

HIGH RISE EDUCATIONAL BUILDING
CENTRAL LONDON
STAGE C COST PLAN

2.0 AREA SCHEDULE

Floor	Open Plan Office	Cellular Offices and Meeting Rooms	Reception	HSC	General Teaching	Education	Labs	LV Switch Room	Copy/ Print Room	Quiet Room	Café and Kitchens	Toilets/ Showers	Changing Rooms	Storage/ Cleaners Cupboard	Circulation	Risers	Stair Core	Lift Core	Balance (Frame, partition, etc)	GROSS INTERNAL AREA
Ground			19	210	300			20	27	10	35	49		43	401	30	16	20	80	1,260
First					97	393			12		21	33		74	279	30	59	20	53	1,070
Second		68				350	342		6			33	35	20	208	30	36	20	69	1,217
Third	226	26				70	454					33		57	184	30	36	20	81	1,217
Fourth		142	20				343			12	12	44	12	32	156	30	36	20	49	908
Fifth	408	108							3	12	29	31		11	105	36	36	20	54	851
Sixth	408	108							3	12	29	31		11	105	36	36	20	54	851
Seventh	408	108							3	12	29	31		11	105	36	36	20	54	851
TOTAL	1,450	560	39	210	397	813	1,139	20	54	58	155	285	47	259	1,543	258	291	160	494	8,225

Notes:
This area schedule is solely for use in the preparation of the cost plan and is not to be used for any other purpose.

Scales for floor plans

Another matter which affects the measurement is the scale of the drawings. This is particularly the case at the beginning when the estimate may be based on a small number of preliminary drawings and sketches. This can be illustrated if we examine the drawings provided in Appendices 1 and 2. They were originally drawn on size A1 paper. The Site Layout (Drawing CP4B/AS/01) is at a scale of 1:100 with landscape orientation. The second drawing (CP4B/AS/02) uses two scales. The Floor Plan and Section A–A use a scale of 1:100 and the two elevations are at 1:200. At A1 size the figured dimensions and the scales are correct.

By examining the drawings at different printed sizes it is possible to see what impact this has on the scales.

A1 The figured dimensions match scales of 1:100 and 1:200.
A2 The figured dimensions on the drawings are at a scale of 1:141 which cannot be used.
A3 The figured dimensions match scales of 1:200 and 1:400.
A4 The figured dimensions on the drawings are a scale of 1:282 which cannot be used.
A5 The figured dimensions match a scale of 1:400 and 1:800.

For this textbook the drawings have been reduced in size to the page size and are reproduced in Appendices 1 and 2.

Although the scales are still shown on the reduced reproductions you will find when you check them that they are no longer accurate. It is not possible to match the figured dimension by taking any figured dimension and measuring it on any common scale. The drawings in the Appendices do not scale accurately for any of the commonly available scales. By enlarging to A1 the scale matches the size of the drawing. It is recommended that an A1 reproduction of each drawing is obtained before starting work on the exercises. By enlarging to A3 it is possible to use scaling with care and at double the scales (1:200 and 1:400) shown on the drawings.

This reduction and enlargement of drawings is common between computer terminals and printers and care needs to be taken to ensure that the scales shown on the drawings are not used to extrapolate dimensions that are not shown as figured dimensions. The preparation of schedules of areas is often done with little information and under severe time constraints. The temptation to use scaling is strong. What this tells us generally, is that we cannot rely on scaling and that figured dimensions on the drawings are the only dimensions that should be used.

Estimates and cost plans

At the estimate stage the designers will normally be trying to maximise the floor area in order to make optimal use of the site. The preparation of accurate and consistent floor areas at this stage provides a common and reliable basis for further variations. After the estimates have been accepted, the design is developed and the cost plans are prepared, based on the revised drawings. At the cost plan stage the important aim is to maintain consistency in the approach to the schedule of areas. There will be value engineering exercises where alternative designs will be tried and costed in order to optimise the design in terms of space and expenditure. This process can radically change both the structure of the building and the areas that are enclosed. A consistent approach to the schedule of areas will enable all parties to understand the changes to both the areas and the costs.

Refurbishments often include a change of use. For example many city centre commercial premises are now being converted to residential accommodation. Table 3.3 shows what this means in terms of how to measure the floor areas. The floor areas would have been measured as the NIA when the building was being used as commercial premises. For the refurbishment and conversion to residential accommodation the floor areas are now measured as the GIFA. To provide transparency to the client both measurements should be included.

For refurbishments, access to occupied areas of the building is often restricted which reduces the extent of physical measurement that is possible. This increases the amount of extrapolation from existing information. The extent to which access is restricted should be included in the report.

Post-contract

The measurement of areas during construction (the post-contract period) for the valuation of progress payments to the contractor requires access which may not be appropriate or possible due to health and safety issues. Although it may not be possible to physically measure the areas, a visual inspection and assessment of areas that have been completed is normal in these circumstances. An assessment of the percentage of the floor area completed will provide sufficient information to value the works.

Variations during construction can provide a change of use of the type of floor area. For example adjustments to ground floor commercial accommodation to provide separate retail areas are common. Again we can refer to Table 3.3 to see what adjustments are necessary. The commercial area would have been built as a shell with no internal partitions. It would be measured using the GIFA. If this area were to be used to provide retail accommodation the measurement would change to the NIA.

Planning constraints may also cause a problem. Social housing can often be a mixture of residential, commercial and retail accommodation. The developers often find that the demand for commercial and retail premises changes during the construction period. For example the demand for nursery facilities to provide for the children in the social housing means that some of the retail units are converted to nurseries which can be classified as commercial premises. This will require changes for the retail areas (measured as NIA) to the commercial areas (measured as GIFA).

Complexity

The Code provides a series of detailed applications that identify the majority of anticipated situations for each of the different types of measurement. There are many different kinds of buildings and new categories continue to emerge. An example of the emergence of new categories is the development from the shop as retail premises to the supermarket and then to the hypermarket or superstore. This means that there are complexities in the Code. For example:

- Domestic houses pay Council Tax which uses GEA (Code, Applications 2, p. 9) and industrial premises pay Rates which use GIA (Code, Applications 6, p. 13).
- New homes use GIA (Code, Applications 8, p. 13) for the estimate of the initial cost of the building, however the cost of the rebuilding for insurance purposes uses GEA (Code, Applications 3, p. 9).

- Ordinary supermarkets use NIA (Code, Applications 10, p. 17) but superstores use GIA (Code, Applications 6, p. 13).
- Residential valuation has developed a series of different measurements that are set out in the Code. This is particularly complex and fails to fulfil the requirements of being comprehensible and consistent. To be useful for the non-expert they require clarification and explanation.
- Existing buildings are not a separate category. For example a warehouse is a non-residential new build premises which is measured before it is built using GIA (Code, Applications 4, p. 13) for the estimate. After they are built they are measured for valuation purposes using GIA (Code, Applications 5, p. 13), for Rates using GIA (Code, Applications 6, p.13) and for insurance purposes for rebuilding, they will probably use GEA (Code, Applications 3, p. 9).
- New build houses are measured before they are built using GIA (Code, Applications 4, p. 13) for the estimate. After they are built they are measured for valuation purposes using GIA (Code, Applications 8, p. 13), for Council Tax using GEA (Code, Applications 2, p. 9) and for insurance purposes for rebuilding GEA (Code, Applications 3, p. 9).
- The need to measure existing buildings therefore only arises for specific purposes.
- Garages and conservatories are measured using GIA (Code, items 2.16 and 12.17, p. 12). Annexes and additions also use GIA but if they are of significantly different construction they should be measured separately (Code, Special use of definitions: Residential, item 28.8, p. 30).

These examples certainly demonstrate the complexity involved. This reflects the variety of buildings and functions and is constantly changing. The professions are familiar with the rules and use them often. However, they may be the cause of some confusion to the client. The Code sets out what the intent should be which is worth repeating '*. . . it does not matter what the professionals may think and understand, it is what the average person thinks and believes that is important in deciding whether the statements are misleading or not.*' (Code of Measuring Practice, 6th edition, 2007, p. 3).

Transparency

The need to maintain accuracy, consistency and clarity is probably best described as transparency, so that the changes are clearly demonstrated. Rather than wait to the end of the project to inform the client of the changes each progress report should provide details of changes to the area schedules.

3.4 GOOD PRACTICE

Areas included and excluded

The junction between different elements of a building can raise questions as to whether to include them in the area schedule. The Code details a series of common interface items that are included and excluded. The examples below illustrate some of these items.

Columns

Columns attached to external walls are included in the GIA floor area (Diagram 3.1).

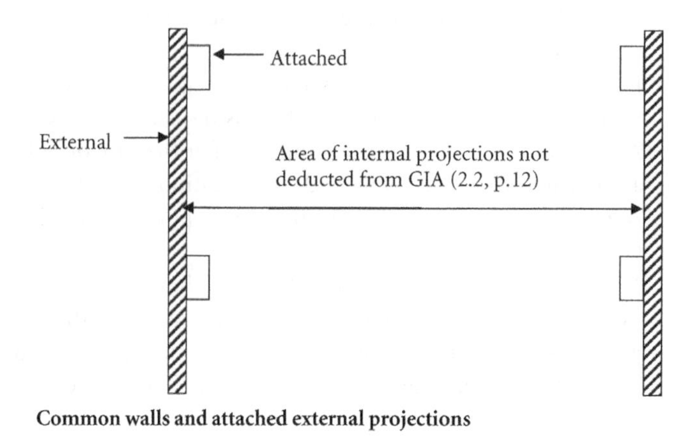

Diagram 3.1 Columns. Source: *The Code of Measuring Practice* 6th edition 2007, p. 14.

Common walls and attached external projections

External walls and their attached projections are included in the measurement of the GEA areas (Diagram 3.2). For common or party walls the centre line is taken. This is highlighted in the dashed line.

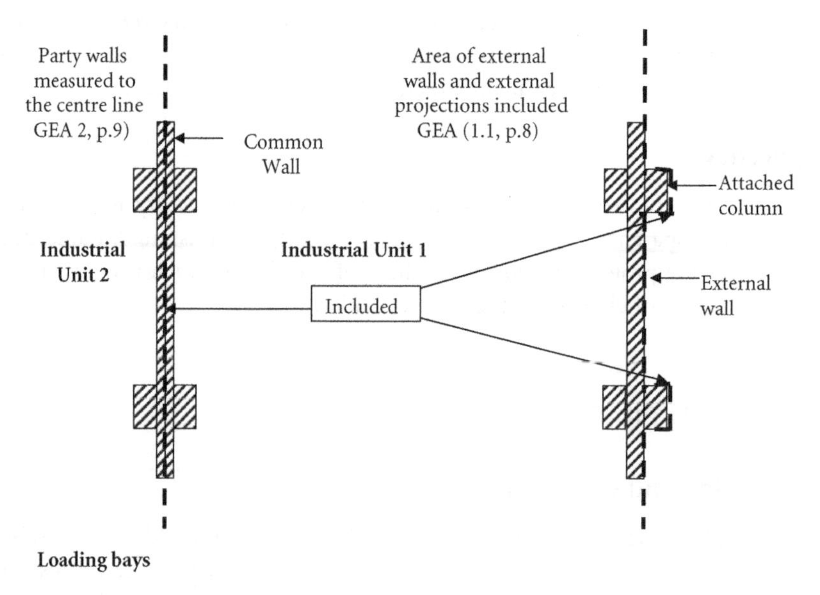

Diagram 3.2 Common walls and attached columns. Source: *The Code of Measuring Practice* 6th edition 2007, p. 10.

Loading bays

Loading bays are included, they are measured (Code, item 2.13, p. 12) but canopies are excluded, they are not measured (Code, item 2.20, p. 12). Diagram D from p. 15 is partially reproduced in Diagram 3.3 and illustrates this item.

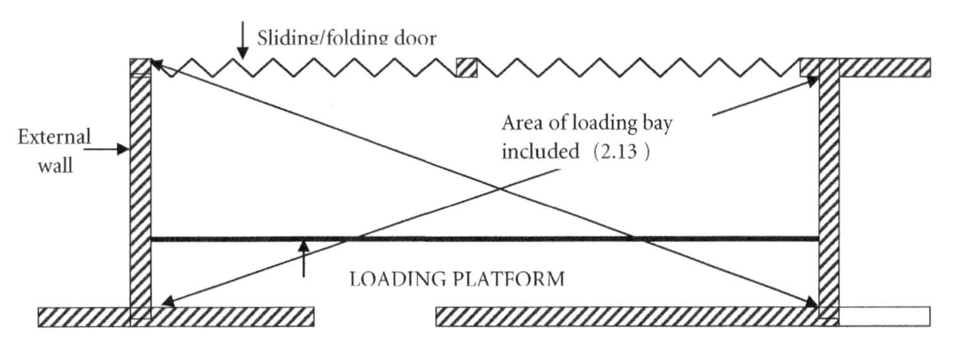

Diagram 3.3 Loading bays.

Canopies

If the loading bay platform is constructed inside the building and the sliding/folding doors move back to the line of the external wall then the former loading area becomes a canopy and is excluded, it is not measured (Code, item 1.17, p. 8; Code, item 2.20, p. 12) as shown in Diagram 3.4.

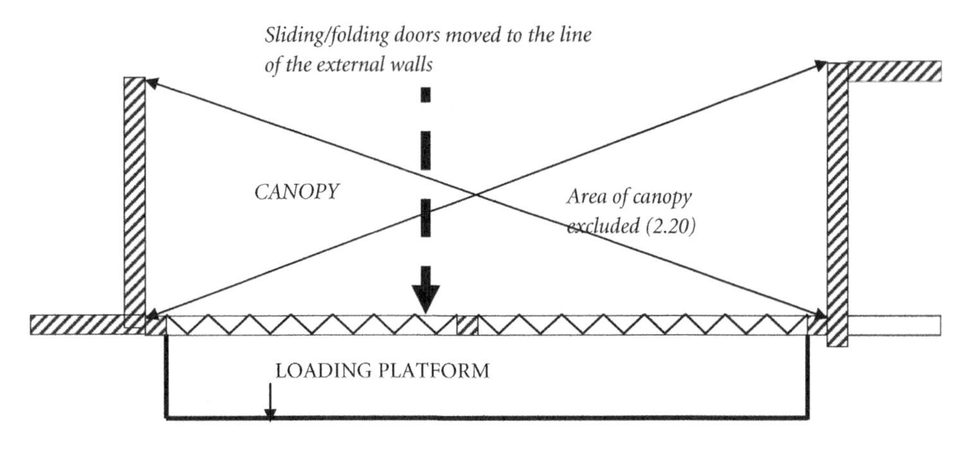

Diagram 3.4 Canopies.

Atria

Atria (Diagram 3.5) are measured separately stating their clear height above the base level (3.0, p. 16). The perennial struggle between aesthetics and financial requirements means that atrium space is often subject to frequent changes. The cost of this open space on each floor is often reduced by filling in one or more of the atrium spaces to create more floor space. For this reason it is best kept as a separate measured item.

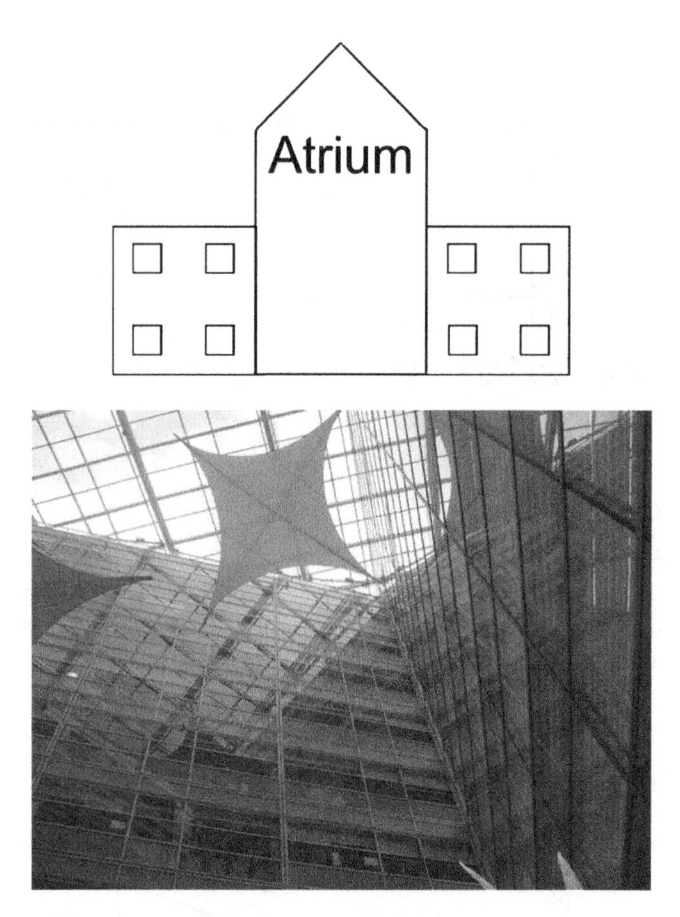

Diagram 3.5 Atria.

Winter gardens

Winter gardens are the provision of a second external wall comprising fully glazed curtain walling where the floor area is extended between the two walls with a balcony and each area is enclosed with glass walls. The balconies have effectively changed from *'external open sided balconies'* (Code, item 2.19, p. 12) which are excluded, to *'internal open-sided balconies'* (Code, item 2.4, p. 12) and are therefore included in the floor area. However, they are not designed as living space but for extensive internal planting and greenery and do not have heating installed. This space is often included in the floor area. Some examples are shown in Diagram 3.6.

Diagram 3.6 Winter gardens.

Specialists

There are specialist consultancies that prepare area schedules. These are often included in reports from the architect, quantity surveyor or project manager. Floor areas are often the subject of vigorous discussion so a report prepared by a third party can have two advantages. First, the expertise is robust because the information comes from a specialist. Secondly, the report is from an objective third party.

Alternative uses

Providing the client with alternative configurations for using the floor space give more potential on how to use the space and allow the architect more freedom on how to create the space. The example below (see Diagram 3.7) uses two different methods of measuring floor areas. Gross internal area (GIA) in accordance with the RICS Code and net useable area (NUA) which does not appear in the Code.

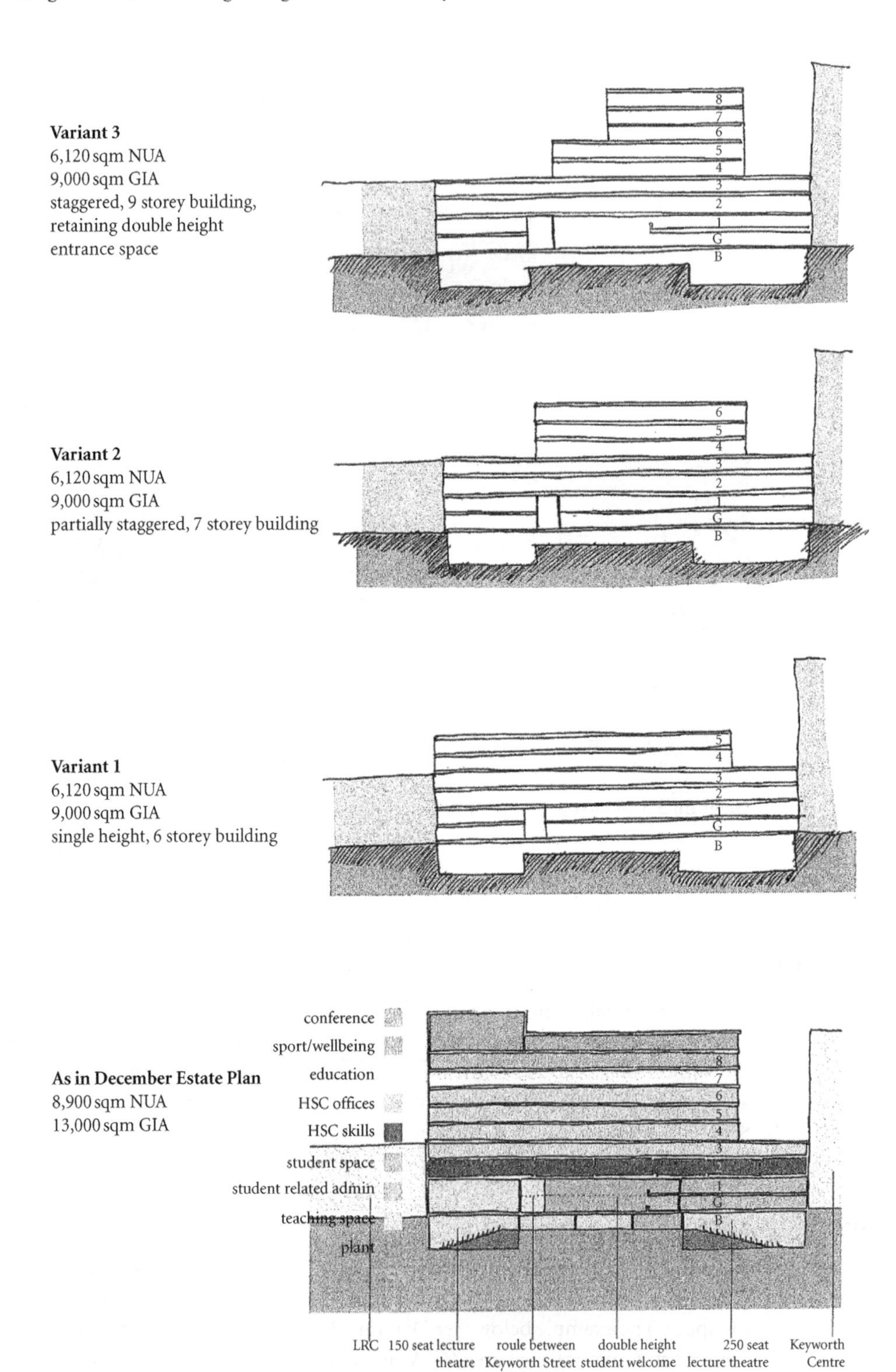

Diagram 3.7 Floor area variables.

What is NUA? It does not appear in the Code. There is no published definition of NUA and it can have variable meanings. This is an example of why the Code can be so useful. Using the standard descriptions removes the need to explain alternative descriptions to confused clients.

Area schedules for estimates and cost plans

Estimate stage

Area schedules increase in complexity as the design develops. At the Estimate stage a simple floor-by-floor area is acceptable. Using the same example one of the floor schedules for the alternative layouts shown in Section 3.5 is set out in Table 3.4.

Table 3.4 Net useable areas (NUA) and net internal areas (NIA).

9th floor	Conference suite		800 m^2	NUA
8th floor	Sport, healthy living and wellbeing		800 m^2	NUA
7th floor	Dept. of Education		800 m^2	NUA
6th floor	HSC faculty offices and meeting rooms		800 m^2	NUA
5th floor	HSC faculty offices and meeting rooms		800 m^2	NUA
4th floor	HSC faculty offices and meeting rooms		800 m^2	NUA
3rd floor	HSC faculty offices and meeting rooms		800 m^2	NUA
2nd floor	HSC faculty skill labs		900 m^2	NUA
1st floor	Student admin + other		600 m^2	NUA
Ground	Student welcome space + other		900 m^2	NUA
Basement	Teaching space + other		900 m^2	NUA
		Total	11,000 m^2	NIA
			13,000 m^2	GIA

The schedule uses NUA, NIA and GIA. The Code gives us the definitions for NIA and GIA, but what is NUA? The use of NUA is an attempt to inform the client of the space that can actually be used on each floor. A summary figure for NIA and GIA is also provided. This additional measurement of the NUA is neither helpful nor transparent because it leads to confusion. The NIA and NUA can be compared as follows:

- NIA. The Code states that NIA excludes areas such as toilets, staircases and corridors. (Code 3.12, 3.14, 3.15, p. 16)
- NUA. In the report that contains this floor schedule useable space is described as an area that '. . . *can be used for any sensible purpose in connection with the purpose for which the premises are to be used*'. So the NUA could include toilets, staircases and corridors.

There is no definition of what constitutes NUA and the difference between NUA and NIA is not provided. This is an example of the confusion that can arise if the standard definitions are not used consistently. Although this area schedule appears to provide useful information the lack of consistency with the Code and the lack of explanation of the difference between NUA, NIA and GIA means that it is confusing and misleading and not in compliance with the Code.

Cost Plan stage

An area schedule as set out in Table 3.5 provides a room-by-room and floor-by-floor analysis. The question of accuracy has been examined earlier with regard to this schedule. Although the measurement of the smallest room may be problematic because such small sizes are difficult to measure the use of a balance number for the area of the internal walls and partitions provides a useful self-adjusting check on the GIA.

Table 3.5 Area schedule.

STAGE C COST PLAN																				
2.0 AREA SCHEDULE																				
Floor	Open Plan Office	Cellular Offices and Meeting Rooms	Reception	HSC	General Teaching	Education	Labs	LV Switch Room	Copy/ Print Room	Quiet Room	Café and Kitchens	Toilets/ Showers	Changing Rooms	Storage/ Cleaners Cupboard	Circulation	Risers	Stair Core	Lift Core	Balance (Frame, partition, etc)	GROSS INTERNAL AREA
Ground			19	210	300			20	27	10	35	49		43	401	30	16	20	80	1,260
First					97	393			12		21	33		74	279	30	59	20	53	1,070
Second		68				350	342		6			33	35	20	208	30	36	20	69	1,217
Third	226	26				70	454					33		57	184	30	36	20	81	1,217
Fourth		142	20				343			12	12	44	12	32	156	30	36	20	49	908
Fifth	408	108							3	12	29	31		11	105	36	36	20	54	851
Sixth	408	108							3	12	29	31		11	105	36	36	20	54	851
Seventh	408	108							3	12	29	31		11	105	36	36	20	54	851
TOTAL	1,450	560	39	210	397	813	1,139	20	54	58	155	285	47	259	1,543	258	291	160	494	8,225

Notes:
This area schedule is solely for use in the preparation of the cost plan and is not to be used for any other purpose.

Cost Plan Stage and Value Engineering

Table 3.6 is an accommodation schedule which allows the developer to compare different types of unit and the amount of accommodation being offered to the purchaser. A comparison of different sized units enables the developer to concentrate the value engineering in areas where the accommodation appears to be too low. The use of square feet (ft^2) is still widely prevalent in residential development and should be provided in area schedules.

Table 3.6 Developer's area schedule.

Date: 05-12-05 (Areas based on drawings issued on 01-12-05)						
NOTE: Areas indicated are Gross Internal						
PLOT No	**BLOCK No**	**CORE**	**LEVEL**	**FLAT TYPE**	**AREA (sq.m)**	**AREA (sq.ft)**
1811	H2	3	1	3 bed	75.77 m^2	815.62 ft^2
1812	H2	3	1	3 bed	81.12 m^2	873.20 ft^2
1813	H2	3	1	2 bed	61.09 m^2	657.55 ft^2
1814	H2	3	1	3 bed	75.86 m^2	816.61 ft^2
1815	H2	3	1	1 bed	51.45 m^2	553.78 ft^2
1816	H2	3	2	3 bed	75.80 m^2	815.94 ft^2
1817	H2	3	2	3 bed	81.17 m^2	873.74 ft^2
1818	H2	3	2	2 bed	61.12 m^2	657.88 ft^2
1819	H2	3	2	3 bed	75.89 m^2	816.93 ft^2
1820	H2	3	2	2 bed	62.43 m^2	672.01 ft^2
1821	H2	3	3	3 bed	75.80 m^2	815.94 ft^2
1822	H2	3	3	3 bed	81.17 m^2	873.74 ft^2
1823	H2	3	3	2 bed	61.12 m^2	857.88 ft^2
1824	H2	3	3	3 bed	75.89 m^2	816.93 ft^2
1825	H2	3	3	2 bed	62.43 m^2	672.01 ft^2
1826	H2	3	4	3 bed	70.83 m^2	762.43 ft^2
1827	H2	3	4	3 bed	80.85 m^2	870.27 ft^2
1828	H2	3	4	2 bed	61.11 m^2	657.86 ft^2
1829	H2	3	4	3 bed	70.71 m^2	761.19 ft^2
1830	H2	3	4	2 bed	66.01 m^2	710.59 ft^2
1837	H2	4	1	1 bed	51.45 m^2	553.77 ft^2

Reproduced by permission of St George PLC.

3.5 PRACTICAL APPLICATION: GIFA LONDON ROAD

An example of the practical application of the Code to measuring the floor area for a commercial office building is provided to allow the procedure to be understood. Appendix 1 is Drawing SDCO/1/1 The Site Layout at a scale of 1:100 as drawn originally on A1 paper. Appendix 2 is Drawing SDCO/1/2. The floor plan and section A–A are at a scale of 1:100 and two elevations at a scale of 1:200, all as drawn originally on A1 paper. Using the rules in the Code the GIFA has been measured on estimating paper (Table 3.7). A query sheet (Table 3.8) is also included. Queries are set down in the query column and then sent to the designer, who will provide the answers. A narrative has been included in italics to provide explanations.

Table 3.7 London Road, GIFA.

London Road

Gross Internal Floor Areas

Drawing Nos. CP4B/AS/01 Floor plan Scale 1:100
CP4B/AS/02 Plan, elevations, section.

Timesing	Dimensions	Squaring	Description	Waste	Notes
				10/4800 48000 ✓	*The figured dimensions are limited.*
			[Bas't-5th	2/6000 12000 ✓	
✓ 7/	48.00 ✓			Conference	*Other dimensions have to be scaled.*
	12.00 ✓	4,032.00		3/4800 14400 ✓	
✓ ✓ 7/½/	19.20 ✓		[Conference	4800 ✓	*A query sheet concerning scaled dimensions should be provided to the architect, as follows.*
				19200 ✓	
	2.00 ✓	134.40	[Exhibition		
	10.50 ✓			2/2750 5500 ✓	
✓ Ddt	10.50 ✓	110.25 ✓		5000 ✓	
				10500 ✓	
4/½/2.75	5.00 ✓				*These dimensions can then be clarified to the benefit of all parties.*
2.75	4.70 ✓		[Scaled		
		23.50 ✓			
		(15.13) ✓		6th Floor	
✓ 3/	4.80 ✓		G/L 8-11		*Checking the dimensions and the calculations at this stage ensures errors can be corrected before they are used elsewhere.*
✓	12.00 ✓	172.80 ✓			
	9.60 ✓			G/L 6/7 2/4800 9600 ✓	
	9.20 ✓	88.32 ✓		6000 ✓	
½/	19.20 ✓			(1200) ✓	
				[Scaled (200) ✓	
	2.00 ✓	19.20 ✓		2/4600 ✓ 9200 ✓	
Scaled] π/	1.70 ✓		[Cupola		
	1.70 ✓	9.08 ✓			
		4,574.42 ✓			

Table 3.8 London Road query sheet.

QUERY SHEET	
LONDON ROAD GIFA	
QUERY	ANSWER
1. Conference grid line dimension GLA+	Scaled at 2000 Confirmed ••.•• As figured dimension
2. Length of hypotenuse for deducts for conference	Scaled at 2700 Confirmed ••.••.••
3. Distance from office to conference	Scaled at 4700 Confirmed ••.••.••
4. Width of 6th floor office	Scaled 6000 Confirmed ••.••.•• Less 1200 + 200 Confirmed ••.••.••
5. Size of bell cupola	Scaled 3400 and circular Confirmed ••.••.••

3.6 SELF-ASSESSMENT EXERCISE: GIFA LONDON ROAD

Using the example shown earlier in this chapter measure the gross internal floor area using the two drawings SDCO/1/1 and 2 included in Appendices 1 and 2, respectively, and the blank estimating paper included in Appendix 3. Rather than simply copying the dimensions from the practical example in Section 3.5, the intention is to begin the practice of measurement by starting with a familiar example to follow. Please prepare your own measurement of the GIFA and also prepare a query sheet of the problems that you have encountered. Then compare it to the example shown in Section 3.5.

Compare your own work with the proposed solution included in Section 3.5.

Self-assess your work on the assessment sheet included in Appendix 4.

The exercises in the following chapters provide self-assessment exercises that deal with different elements than the worked example in the practical application sections. For example in Chapter 4 the practical example measures the frame and the self-assessment exercise measures the basement. However, all the worked examples and the self-assessment exercises use the same set of drawings which should enable familiarity with the vocabulary and the technology to be improved as each exercise is progressively undertaken.

To provide further assistance there are dedicated websites at http://ostrowskiquantities.com and at Wiley Blackwell (http://www.wiley.com/go/ostrowski/estimating). It is hoped that the provision of this will go some way towards explaining the concepts and principles more clearly than using the printed word alone.

4 How to Use the New Rules of Measurement 1

4.1 INTRODUCTION

The purpose of NRM 1

The purpose of the NRM 1 is set out in the Introduction. This includes a '. . . *standard set of measurement rules* . . .' and '. . . *effective and accurate cost advice* . . .' (p. 2). NRM 1 introduces two new standard methods of measurement for quantities in the early stages of the development of a construction project. One for estimates and one for cost plans. The NRM also provides the information concerning how to measure floor areas in a series of appendices. This replicates most of the Code of Measuring Practice, 6th edition, 2007.

The common themes of standardisation, accuracy and consistency have emerged in these two published documents. Standardisation is provided by setting out a schedule of elements to measure and a matrix of how to measure them. Accuracy is provided by precise definitions of dimensions to the nearest millimetre and quantities to two decimal points of a metre. Consistency comes from a series of carefully defined circumstances when particular methods of measurement should be used.

Estimating and Cost Planning Using the New Rules of Measurement, First Edition. Sean D.C. Ostrowski.

Estimates and cost plans are probably the most important service that quantity surveyors provide for their clients. The provision of accurate and prompt estimates enables the developer to decide on the financial strength of the development when these costs are incorporated into a feasibility exercise. The first published anticipated cost in an estimate is often the most significant figure of the whole development. As the design evolves the estimates become more detailed cost plans which then provide a more accurate anticipated tender price.

The accuracy of these financial exercises is dependent on the expertise and experience of the quantity surveyor and the extent of information that is provided by the design team. The professional practices that provide this service build up this expertise, develop a measurement and cost database and a form of presentation that provides a unique service to the client and which emphasises the particular strengths of the practice in this area of development.

The problems that the client encounters are twofold. First, each practice prepares and presents the estimates and cost plans to match their particular expertise and procedures. Each practice will measure and price the work differently. The possibility of comparing anticipated costs between different practices or different contracts is made more difficult by the reports not measuring areas and quantities on the same basis. This means that the costing of the various elements of work are not comparable.

Secondly, the changes made to the costs at the various stages of development are often difficult to follow. The changes from the estimate to the cost plan, then to the tender and to the final account are often obscured by alterations in the methods of measurement and pricing as the design and construction proceeds. This means that the transparency that the client desires is not provided in the audit trail of information that demonstrates the changes to the design and their financial consequences.

The provision of a set of rules, the NRM for estimates and costs plans, which standardises the way estimates and cost plans are measured is a considerable step towards improving the service to the client.

Practice and procedure

We commence our examination of the NRM by looking briefly at the practice, procedure and different standard methods of measurement for estimates and cost plans. Each will be examined in more detail in subsequent chapters. The constituent parts and the elements that are built up in each constituent are explained. Each element is subject to a hierarchical system of levels and reference numbers and examples are provided to demonstrate how they work.

There are different methods of measurements for both estimates and cost plans, each of which has specific and detailed rules. The NRM has provided a set of rules which is both prescriptive and detailed. Although it may appear to be rather technical and complex at this stage, a preliminary overview indicates that there is structure and precision to the methods of measurement. The example below compares the procedure for estimates and cost plans of basements and allows the pages, columns, units of measurement and references to be identified by reference to the NRM.

For basement estimates the procedure is as follows:

- p. 27 Table 2.1 provides the rules
- Column 1 provides the group element, substructure

- Column 2 the element, substructure
- Column 3 the unit of measurement which is square metres
- Column 4 describes basements at item 3
- The references is item 1.1.3

For basement cost plans the procedure is as follows:

- NRM, p. 74, Part 4 provides the tabulated rules of measurement for elemental cost planning
- p. 89 provides the group element, item 1 substructure
- p. 89 provides the element, item 1.1 substructure
- p. 98 column 1 provides the sub-element, item 1.1.4 basement excavation
- p. 98 column 2 provides the component, item 1.1.4.1 basement excavation
- p. 98 column 3 the unit of measurement which is cubic metres
- The reference is item 1.1.4.1
- Disposal of excavated material is also measureable as item 1.1.4.2 in cubic metres
- Earthwork support is also measureable as item 1.1.4.4 in square metres

The information requirements for estimates and cost plans are introduced at this stage as they are an integral part of NRM 1.

An example of the practical application of NRM 1 concerning items that are included and excluded is provided to allow the procedure to be understood. A narrative has been included in italics to provide explanations. A self-assessment exercise is provided to allow practice of the layout of the constituents and elements of an estimate.

4.2 FRAMEWORK

Work stages

One of the first things to establish is when the estimates and cost plans are prepared in the development programme. The order for this is set out in NRM 1, p. 8 and Table 4.1 is similar. The quantity surveyor is particularly concerned with the middle column which concerns the cost of the project. However, the other members of the design team have their own programme with different priorities. The intent is to correlate estimates and cost plans with the RIBA work stages and the Office of Government Commerce (OGC) Gateways. The three processes shown below concern the different elements of design, finance and construction. The correlation cannot be exact, and so there is some overlap of the stages. Because the stages do not correspond and because they overlap, the concept of freezing the design in order to fix the budget or obtain approval is rarely appropriate or workable.

Table 4.1 Planning stages in NRM.

	RIBA Work Stages	RICS formal cost estimating and elemental cost planning stages	OGC Gateways (Applicable to building projects)
Preparation	A Approval	Order of cost estimate	1 Business Justification
	B Design Brief		2 Delivery Strategy
Design	C Concept	Formal Cost Plan 1	3A Design Brief and Concept Approval
	D Design Development	Formal Cost Plan 2	
	E Technical Design	Formal Cost Plan 3	
Pre-Construction	F Production Information	Pre-tender estimate	3B Detail Design Approval
	G Tender Information		
	H Tender Action	Post-tender estimate	3C Investment Decision
Construction	J Mobilisation		
	K Construction to Practical Completion		4 Readiness for Service
Use	L Post Practical Completion		5 Operations Review and Benefits Realisation

Source: RICS (2012), p. 8. Reproduced by permission of the RICS.

Constituents

There are several constituents of an estimate and they are set out in NRM 1, p. 22 and summarised in Table 4.2. We provide a brief review here which introduces each constituent. A more detailed examination will be given in Chapter 5.

Table 4.2 Constituents of an estimate.

Constituents of estimates
Building works
Main contractor's preliminaries
Main contractor's overhead and profit
Project/design team fees
Other development/project costs
Base cost estimate
Risk allowances
Inflation
Cost limit
VAT assessment

Source: RICS (2012), p. 22. Reproduced by permission of the RICS.

Building works estimates

The first constituent is the building works estimate. This is the cost of the building works elements numbered 0–8 in NRM, pp. 24–5. The estimate comprises a series of standard elements which is measured using NRM 1 for the building works and floor areas that are the gross internal floor area (GIFA). When applied to each stage of the process and to different projects this approach provides a consistent and accurate methodology which enables the development of the design and the cost of the works to be followed from inception to completion.

There are alternative methods of measurement. They are, for example, the cost of a commercial building per office work station or of a hospital per bed. They can only be prepared if the quantity surveyor has considerable expertise and experience in that particular kind of contract. They are not considered further in this textbook. Functional units are briefly described in NRM 1, p. 23 and in Appendix B of the NRM.

In addition to the elemental arrangement of the building works the NRM adds the following constituents.

Main contractor's preliminaries

The main contractor's preliminaries are the costs of the site requirements, including management, accommodation and plant. They are described in NRM 1, p. 38 for estimates and listed in group element 9 in NRM, p. 277 for cost plans. Preliminaries, risk, overheads and profit will be considered in detail in Chapter 8.

Main contractor's overheads and profit

The main contractor's overheads and profit are described in NRM 1, p. 39 for estimates and are separate items in group element 10 in NRM 1, p. 307 for cost plans.

Design fees

The NRM provides a schedule of design costs for the project team which are described in NRM 1, p. 40 for estimates and scheduled in group element 11 in NRM 1, pp. 309–12 for cost plans.

Other development costs

These costs include planning costs, insurances and finance charges are described on NRM 1 p.40 for estimates and are detailed in group element 12 in pp. 317–20 for cost plans.

Base cost estimate

The measurable cost of the labour and materials, preliminaries, overheads and profit, fees and other development costs.

Risks

The NRM provides four risk items and each requires a detailed calculation. These risk items take the place of the 'contingency' which used to provide a simple overall risk item with no analysis. They are described on pp. 41–3 for estimates and detailed in group element 13 in NRM 1, pp. 321–5 for cost plans.

Inflation

This is an assessment of the inflation that is likely to be incurred both during the pre-contract period and the construction period, the post-contract period. These costs are described on pp. 43–5 for estimates and group element 14 on p. 327 for cost plans.

Cost limit

The cost of the labour and materials, preliminaries, overheads and profit, fees, other development costs, risk and inflation. The intention is plain. It is to provide the client with some reassurance that the final cost is the limit of the expenditure.

Value Added Tax

The NRM suggests that Value Added Tax (VAT) is excluded from the estimate because it is complex. However, it is a significant area of expenditure and a comprehensive estimate of costs is not complete without a VAT assessment.

An alternative report layout showing the estimate of each of these elements separately is provided in Table 4.3.

Table 4.3 Estimate: building works.

Estimate

			£	£/m²
1	**Substructure**			
2	**Superstructure**			
	2.1	Frame		
	2.2	Upper floors		
	2.3	Roof		
	2.4	Stairs		
	2.5	External walls		
	2.6	Windows and external doors		
	2.7	Internal walls and partitions		
	2.8	Internal doors		
3	**Internal finishes**			
	3.1	Wall finishes		
	3.2	Floor finishes		
	3.3	Ceiling finishes		
4	**Fittings and furnishings**			
5	**Services**			
	5.1	Sanitary installation		
	5.2	Services equipment		
	5.3	Disposal installation		
	5.4	Water installation		
	5.5	Heat source		
	5.6	Space heating and air conditioning		
	5.7	Ventilation systems		
	5.8	Electrical installation		
	5.9	Gas installations		
	5.10	Lift installations		
	5.11	Fire and lightning protection		
	5.12	Communications systems		
	5.13	Specialist installations		
	5.14	Builders' work in connection		
	5.15	Test and commission		
		carried forward		

Table 4.3 (*Continued*)

Estimate

			£	£/m^2
		brought forward		
6	**Complete buildings**			
7	**Work to existing**			
8	**External works**			
9	**Facilitating works**			
10	**Preliminaries**			
11	**Overheads and profit**			
12	**Fees**			
13	**Other costs**			
14	**Risk**			
15	**Inflation**	**Cost limit £**		**/m^2**
				/ft^2
16	**Exclusions** 1. Value Added Tax 2. Work outside perimeter of building			

4.3 ESTIMATES

Elements

The anticipated cost of a building is calculated as the total cost of each of the elements of the building. For estimates the 'condensed' list of elements is used as shown in NRM 1, p. 24. This condensed list for estimates comprises two levels of elements. The first level is of Group Elements and the second level of Elements. It is progressively expanded into five levels for cost plans as more information becomes available. (The condensed list is referred to as part of the cost plan 1 described on p. 54, section 3.11.4.iv). The NRM adds a new item for Element 0 entitled 'Facilitating works'. The condensed list of elements is partially reproduced in Table 4.4.

Table 4.4 Levels of elements for estimates.

LEVEL 1	LEVEL 2
Group element	**Element**
1. Substructure	
2. Superstructure	2.1 Frame
	2.2 Upper floors
	2.3 Roof
	2.4 Stairs and ramps
	2.5 External walls
	2.6 Windows and external doors
	2.7 Internal walls and partitions
etc.	etc.

Source: RICS (2012), pp. 24–5, 54. Reproduced by permission of the RICS.

Methods of measurement for estimates

At this point it is appropriate to introduce the first major development introduced by the NRM. The method of measurement for estimates is a new standard method of measurement. It is only for use on estimates. It uses primarily superficial areas based on the GIFA. There is another new standard method of measurement for cost plans. The method of measurement for estimates is set out in Table 2.1 (NRM 1, pp. 27–36). An example for substructure is described below and a more examples are described in detail in Chapter 5 of this textbook.

Substructure (NRM, p. 27)

The measurement is superficial in square metres. The rules clearly indicate that the ground or lowest floor slab is part of the substructure and is measured in square metres not cubic metres as shown in Table 4.5. How to convert the normal cubic measurement for this element into a new compound item for the new superficial measurement is explained in Chapters 2, 5 and 6.

Table 4.5 Substructure estimates.

Group element	Element	Unit	Measurement rules
1 Substructure		m^2	1 The area measured is the area of the lowest floor measured to the internal face of the external perimeter walls. 2 The area of the lowest floor shall be measured in accordance with the rules of measurement for ascertaining the gross internal floor area (GIFA). 3 Areas of basement to be shown separately. 4 The area of basement shall be measured in accordance with the rules of measurement for ascertaining the GIFA.

Source: RICS (2012), p. 27. Reproduced by permission of the RICS.

4.4 COST PLANS

CP1, CP2, CP3

As the development progresses past the design brief and the scheme becomes a reality, the design is developed and the financial reports can reflect more and better information. These reports are a series of cost plans and the NRM sets out three stages of progressive cost plans as shown on pp. 54–5.

CP1

CP1 uses a condensed list of two levels of elements and a new method of measurement. The condensed list of elements is the same as those used in estimates. However the new method of measurement is different.

CP2

CP2 is based on the completion of the design development. It is '. . . *developed by cost checking of cost significant cost targets for elements* . . .' (p. 55). It uses an expanded list of elements using three to five element levels. There is no further measurement except for the changes from CP1 that require new measurement.

CP3

CP3) is based on '. . . *technical designs, specifications and detailed information for construction* . . .' (p. 55). It is a progression of CP2 and is also to be used for appraising tenders.

These three stages will be examined in more detail in Chapter 6.

Elements and levels

The anticipated cost of a building is calculated as the total cost of each of the elements of the building. It is progressively expanded into five levels of elements for cost plans as more information becomes available. The relationship between elements and levels is that the elements describe the structure and components of the building and the levels describe how to measure them.

The first cost plan, CP1, uses a condensed list of two levels of elements described in NRM, p. 54, section 3.11.4.iv. If the information is available CP1 can include further levels although it is condensed to comprise only two levels in the published cost plan.

The elements for subsequent cost plans are expanded initially into three levels as Appendix E in NRM, p. 347. They are level 1 Group element; level 2 Element; level 3 Sub-element. An extract of Appendix E is shown in Table 4.6 for the first three levels.

Table 4.6 Levels of elements for cost planning.

LEVEL 1 **Group element**	**LEVEL 2** **Element**	**LEVEL 3** **Sub-element**
2 Superstructure	2.1 Frame	1 Steel frames
	2.2 Upper floors	2 Space decks
	2.3 Roof	3 Concrete casings to steel frames
	2.4 Stairs and ramps	4 Concrete frames
	2.5 External walls	5 Timber frames
	2.6 Windows and external doors	6 Specialist frames
etc.	2.7 Internal doors	etc.

Source: RICS (2012), p. 347. Reproduced by permission of the RICS.

For the subsequent cost plans (CP2 and CP3) the elements can be expanded to five levels as shown in Tables 4.7 and 4.8 for steelwork in Part 4: Tabulated rules of measurement for elemental cost planning in NRM 1, p. 75. The intent is to allow the measurement details to expand within a standard, accurate and consistent framework. Level 5 can only be used if sufficient information is provided concerning the subcomponents, the fixtures and fittings.

Table 4.7 The NRM and levels.

LEVEL 1 Group element 2: Superstructure

Group element 2 comprises the following elements:

2.1 Frame

2.2 etc.

LEVEL 2 Element 2.1: Frame

LEVEL 3 Subelement	LEVEL 4 Component	Unit	LEVEL 5 Measurement rules for comparisom	Included	Excluded
1 Steel frames **Definitions** Structural steelwork in frames, including all components, fittings, fixings and components.	1 Structural steel frame, including fittings and fixings: details, including size of column grid (m), to be stated.	t	C1 The total mass of the steel frame is to be stated. The mass of framing includes all fittings and components.	1 Structural steel frame etc.	1 Space frames and decks etc.
etc.	etc.	etc.	etc.		

Source: RICS (2012), p. 103. Reproduced by permission of the RICS.

Table 4.8 Element levels for steel frame.

LEVEL	ELEMENT	Ref.	PAGE	UNIT	DESCRIPTION
1	Group element	2	103		Superstructure
2	Element	2.1			Frame
3	Sub-element	2.1.1			Steel frame
4	Component	2.1.1.1		t	Structural steel frame . . . etc.
5	Sub-component	2.1.1.1.C1			Total mass of steel frame . . . etc.

Method of measurement for cost plans

The method of measurement for cost plans is a new standard method of measurement. It is for use on cost plans only and is the largest section of the NRM, spanning pages 74–328 and is set out in Part 4: Tabulated rules of measurement for elemental cost planning. To introduce this method of measurement an example is described below for the substructure basement excavation and a more detailed explanation is included in Chapter 5 of this textbook.

Table 4.9 Basement method of measurement.

Element 1.1: Substructure

Sub-element	Component	Unit	Measurement rules for components	Included	Excluded
5 Basement excavation **Definitions** Bulk excavation required for construction of floors below ground level.	1 Basement excavation: details, including average depth of excavation, to be stated. 2 Disposal of excavated material: details to be stated. 3 Extra. . . . etc. 4 Earthwork support: details to be stated.	m^3 m^2	C1 The volume measured for basement excavation . . . and the like.	1 Bulk excavation to form basements and the like.	1 Excavation and earthwork . . . etc.
etc.	etc.	etc.	etc.	etc.	

Source: RICS (2012), p. 65. Reproduced by permission of the RICS.

4.5 INFORMATION

One of the major innovations that NRM 1 introduces is the detailed schedules of information to be provided by the design team for the preparation of the estimates and the cost plans. The schedules provide a clear indication of what information is to be available at each stage. The informal arrangement of providing estimates based on whatever information is available has now been replaced with formal stages for information release which will then allow the appropriate estimate or cost plan to be prepared.

This discipline increases the coordination between the designers and will highlight to the client the areas where information remains outstanding. The acceptance and implementation of these information schedules by the design team as stages in the design process that need to be completed is one of the major challenges of the NRM.

A brief review of the information requirements at this stage will complete the review on how to use the NRM. A more detailed appraisal of the information schedules will follow in Chapter 7.

Information required for the estimate (NRM, p. 20, section 2.3)

Information required from the client
This schedule applies once the site has been acquired.

Information required from the architect
The provision of several of these items is unlikely at the estimate stage.

Information required from the structural and services engineers
The provision of the indicative structural layout is one of the most significant elements of the estimate and should be added to section 2.3.4

The information listed in this schedule is comprehensive and assumes that the site has been purchased. The first approximate estimate before the site has been purchased remains informal and will be the report that enables the site to be purchased and the reported figure in this initial, preliminary, approximate will be the most enduring. The problem remains the provision of information for this initial estimate.

Information required for the cost plans
(NRM 1, p. 53, section 3.8) and Appendix F (NRM, p. 352)

These schedules provide a progressive and complete schedule of all the information required to construct the building. Construction drawings or Detailed Design Approval Office of Government Commerce (OGC) Gateway 3B is part of the preparation for the tender documentation. This includes the bills of quantities or schedule of works or work packages and the contract documentation.

The preparation of these documents is part of the final design stage to finalise the construction documents. Much of this information will only become available at RIBA Work Stage F Production Information and can be used for pre-tender estimates but will not be available in time to prepare the final cost plans.

4.6 PRACTICAL APPLICATION: INCLUDED AND EXCLUDED

NRM 1, Table 4 includes columns which describes items that are included and excluded. For example NRM 1, pp. 105–6 deals with openings to concrete walls (Table 4.10).

Table 4.10 Included and excluded: openings.

Element 2.1: Frame

Sub-element	Component	Unit	Measurement rules for components	Included	Excluded
4 Concrete frames **Definitions** Concrete columns and beams.	4. Extra over walls for forming openings in walls for doors, windows, screens and the like: details, including thickness of wall (mm), overall size of opening (m) and type of formwork finish, to be stated.	nr	C4 Walls: the area measured is the area of the wall, measured on the centre line of the wall. No deduction is made for door openings, windows, screens and the like.	13 Forming openings for doors, windows, screens and the like.	

Source: RICS (2012), pp. 105 and 106. Reproduced by permission of the RICS.

The meaning of these separate entries in different columns is that the walls are measured gross, over all openings, in m^2 and a separate measure is added or ‘included’ for the openings which are enumerated.

A further example in NRM, p. 105–6 deals with designed joints (Table 4.11).

Table 4.11 Included and excluded: designed joints.

Element 2.1: Frame

Sub-element	Component	Unit	Measurement rules for components	Included	Excluded
4 Concrete frames **Definitions** Concrete columns and beams.	5 Designed joints: details to be stated.	m		9 Designed joints (eg to walls).	

Source: RICS (2012), p. 105–6. Reproduced by permission of the RICS.

The meaning of these separate entries in different columns is that the designed joints are a separate measure in linear metres which is added or 'included'. Similar examples appear in the NRM. Although these items appear to be ambiguous the intention is that items in the included column are to be measured.

4.7 SELF-ASSESSMENT EXERCISE: CONVERSION TO NRM

An example of a typical estimate is given in Table 4.12. This does not use the layout of the NRM. This exercise is to establish what is necessary to convert it to the NRM format. Probably the easiest way to do this is to create a spreadsheet in the NRM format and convert the items in the estimate below into a NRM estimate. In addition, you should prepare a query sheet and compare the specification provided with the information schedules in the NRM.

Compare your own work with the proposed solution included in Appendix 6 (Appendix 6 is on the website (http://www.wiley.com/go/ostrowski/estimating)).

Self-assess your work on the assessment sheet included in Appendix 4 (Appendix 4 is on the website (http://www.wiley.com/go/ostrowski/estimating)).

To provide further assistance there are dedicated websites at http://ostrowskiquantities.com and at Wiley Blackwell (http://www.wiley.com/go/ostrowski/estimating). It is hoped that the provision of this will go some way towards explaining the concepts and principles more clearly than using the printed word alone.

Table 4.12 Typical estimate.

Element	£'000	Brief specification of works
Substructure	1,300	Piled foundations Reinforced concrete basement
Frame	700	Steel on concrete
Upper floors	600	Hollow rib and concrete or precast
Roof	250	Flat asphalt covered plus rooflights
Stairs	200	Steel or reinforced concrete
External cladding	2,400	Proprietary system with 40% opening windows
Internal walls/ partitions	1,200	Brickwork core walls Laminated faced WC cubicles Veneered doors Offices (50 No.)
Wall finishes	300	Plaster Tiling to WCs
Floor finishes	600	150 mm raised floor Carpet/vinyl
Ceiling finishes	350	Mineral fibre tiles suspended ceilings

(Continued)

Table 4.12 (*Continued*)

Element	£'000	Brief specification of works	
Decorations	350	Paint	
Fittings + fixtures	600		£'000
		Basement teaching space	200
		Ground floor reception	100
		Shelving etc.	150
		Tea points etc.	150
		NB excludes furniture + equipment	
Sanitary and plumbing	400	Services/WCs/tea points White sanitary goods	
Mechanical installations	1,500	Air conditioning to basement lecture theatres/conference Air conditioning to communications room Central heating elsewhere Ventilation to WCs BMS controls Builders' work	
Electrical installations	1,750	Mains power + distribution Electrics to mechanical plant Underfloor power track + outlets Office and special lighting Builders work	
Life safety	250	Fire alarms Dry riser NB: assume sprinkler not required	
Special installation	900		£'000
		CAT 6 data cabling	350
		CCTV	200
		Card access system	150
		Audio/visual/hearing loops etc.	200
Lifts	750	5 No. lifts (incl goods) to serve all floors	
Utilities	250	Connecting gas, water, electricity	
External works	300	Provisional allowance	
Preliminaries	1,750	Contractor's on-site costs for a 16–18 month programme	
Contingencies	800	5% construction contingencies	
	£17,500	**or £1,950/m² GFA**	

5 NRM 1 Estimates

5.1 PRACTICE AND PROCEDURE

Introduction

In this chapter we examine how to prepare estimates using the NRM which includes a new standard method of measurement. The procedures of building up the constituents and elements are elaborated which leads to the provision of an estimate with a cost limit. The specific method of measurement for estimates is examined in detail and examples of different elements are provided. The use of compound items is described to enable orthodox methods of measurement to become compliant with the requirements of NRM 1. The practical example in Section 5.3 allows the procedure to be followed for the measurement of a basement estimate and the self-assessment exercise in Section 5.4 allows practice to be undertaken for the measurement of the frame element of an estimate.

Stages

The order for estimates corresponds to RIBA Stage A Appraisal and Stage B Design Brief and to Office of Government Commerce (OGC) Gateways Stage 1 (Justification) and part of Stage 2 (Delivery Strategy). This is set out in Table 5.1.

Estimating and Cost Planning Using the New Rules of Measurement, First Edition. Sean D.C. Ostrowski.

Table 5.1 Work Stages for estimates.

<table>
<tr><th colspan="2">RIBA Work Stages</th><th>RICS formal cost estimating and elemental cost planning stages</th><th>OGC Gateways (Applicable to building projects)</th></tr>
<tr><td rowspan="2">Preparation</td><td>A Approval</td><td rowspan="2">Order of cost estimate</td><td>1 Business Justification</td></tr>
<tr><td>B Design Brief</td><td>2 Delivery Strategy</td></tr>
<tr><td></td><td></td><td></td><td>3A Design Brief and Concept Approval</td></tr>
</table>

Source: RICS (2012, p. 8).

The intent is to correlate estimates with the RIBA Work Stages and the OGC Gateways. The three processes shown in Table 5.1 concern the different elements of design, finance and construction. This means that the correlation cannot be exact and so there is some overlap of the stages.

There is often an earlier stage prior to Approval and the function of this initial estimate is to enable the scheme to go ahead. It usually predates the purchase of the land and the financial arrangements. Some estimates do not show a positive financial situation and will result in the decision that the scheme in its current layout cannot proceed. This would indicate that the estimate at this particular stage, commonly called the approximate or preliminary estimate, correlates most closely to the OGC Business Justification stage, and is an earlier stage estimate that has no overlap with the Design Brief and the Delivery Strategy.

Constituents

The constituents of an estimate are set out in Table 5.2 and will be examined in turn.

Table 5.2 Constituents of an estimate.

Constituents of estimates
Building works (1)
Main contractor's preliminaries (2)
Sub-total (3) [(3) = (1) + (2)]
Main contractor's overhead and profit (4)
Works cost (5) [(5) = (3) + (4)]
Project/design team fees (6)
Sub-total (7) [(7) = (5) + (6)]
Other development/project costs (8)
Base cost estimate (9) [(9) = (7) + (8)]

Table 5.2 (*Continued*)

Constituents of estimates	
Risk allowances [10] (a) Design development risks [10a] (b) Construction risks [10b] (c) Employer change risks [10c] (d) Employer other risks [10d]	[(10) = (10a) + (10b) + (10c) + (10d)]
	Cost limit [11] [(11) = (9) + (10)]
Tender inflation[12]	
	Cost limit [13] [(13) = (11) + (12)]
Construction inflation[14]	
	Cost limit [15] [(15) = (13) + (14)]
VAT assessment	

Source: RICS (2012, p. 22). Reproduced by permission of the RICS.

Building works estimate[1]

This comprises group elements 0–8 for the building works (see Table 5.3). This condensed list comprises the first level of group elements and the second level of elements. This work is to be measured in accordance with NRM 1.

Table 5.3 Elements for the building works part of an estimate.

Group element	Element
0 Facilitating works	
1 Substructure	
2 Superstructure	2.1 Frame
	2.2 Upper floors
	2.3 Roof
	2.4 Stairs and ramps
	2.5 External walls
	2.6 Windows and external doors
	2.7 Internal walls and partitions
	2.8 Internal doors
3 Internal finishes	3.1 Wall finishes
	3.2 Floor finishes
	3.3 Ceiling finishes

(*Continued*)

Table 5.3 (*Continued*)

Group element	Element
4 Fittings, furnishings and equipment	
5 Services	5.1 Sanitary appliances
	5.2 Services equipment
	5.3 Disposal installations
	5.4 Water installations
	5.5 Heat source
	5.6 Space heating and air conditioning
	5.7 Ventilation
	5.8 Electrical installations
	5.9 Fuel installations
	5.10 Lift and conveyor installations
	5.11 Fire and lightning protection
	5.12 Communications, security and control systems
	5.13 Special installations
	5.14 Builders' work in connection with services
6 Prefabricated buildings	
7 Works to existing buildings	
8 External works	

Source: RICS (2012, p. 24–5). Reproduced by permission of the RICS.

The constituents continue to be built up by adding further group elements as follows.

Group element 9: Main Contractor's Preliminaries[2]

(See NRM, p. 36.) The calculation required for main contractor's preliminaries in an estimate is shown in item 2.11, NRM 1, p. 38. They are simplified to a percentage of the cost of the works. However, the best way of getting an accurate estimate of the preliminaries is to measure them. They will be examined in detail in Chapter 8 of this textbook.

Group element 10: Main Contractor's Overheads and Profit[4]

(See NRM, p. 36.) The calculation required for the main contractor's overheads and profit is item 2.11, NRM, p. 39. It is a percentage of the cost of the works. They are shown as a single entry in the summary. However, both are significant items in the tendering process with different factors applicable to both and they should be shown separately. They are considered in Chapter 8.

Group element 11: Design Fees[6]

The estimate for design fees is item 2.13 (NRM 1, p. 40) and is an allowance, usually a percentage. The detailed description of what is included in this constituent is found in Cost Plan Group Element 11 (NRM 1, p. 309).

Group element 12: Other Development Costs[8]

The estimate for other development costs is item 2.14 (NRM 1, p. 40) and is an allowance usually a percentage. The detailed description of what is included in this constituent is found in the cost plan group element 12 (NRM 1, p. 317). This includes a series of fees in addition to design fees. It also includes the consequences of planning gain, insurances and fittings and furniture. There is a risk of confusion between group element 4 'Fittings, furniture and equipment' and element 12.1.10 'Fittings, furnishings and equipment' in NRM 1, p. 318. The cost should be measured in one element only. These items can add considerably to the direct costs of the building works and are usually excluded because the information is unlikely to be available.

Base Cost Estimate[9]

Taken together, elements 0–12 are the base cost estimate. This is the cost of the works that can be ascertained by measurement and calculation and includes items that are direct costs in addition to building works.

The final two group elements are anticipated estimates of costs that might be incurred in the future.

Risks[10]

These costs are item 2.15 in NRM 1, p. 41 (Cost plan group element 13, NRM 1, p. 321). There are four main risk items and each requires a detailed calculation. Design development risk, the anticipated problems of ensuring the design is compliant with the requirements of the quality and budget. The construction risk is the anticipated problems of construction to time and budget. The employer change risk anticipates the extent to which the employer will generate design changes that will impact on the budget and the programme. The final risk item is the Employer's Other Risks, a category which tries to anticipate problems that the employer might have with end-user requirements, the programme, finance and the impact of third parties like employer's agents and project managers. These items provide a challenge to the quantity surveyor. The relative certainties of measurement and pricing are now superseded by the need for a series of professional judgements which highlight the most problematical areas of the construction process. At the estimate stage these allowances will be approximations. Risk is examined in detail in Chapter 8.

Inflation[12]

These costs are item 2.16 in NRM 1, p. 43 (Cost plan group element 14, NRM 1, p. 327). They are an assessment of the inflation that is likely to be incurred during both the pre-contract and the post-contract periods and are percentages.

Cost Limit[15]

Taken together, elements 0–15 are the cost limit. This is the anticipated maximum cost of the works and includes an element for risk and inflation. The NRM describes the final cost as the cost limit. The intention is plain. It is to provide the client with some reassurance that the final cost is the limit of the expenditure. If all the items that have been described in the elements are included, the cost will be the maximum cost. However, many of these costs require considerable information and a substantial amount of this information is unlikely to be available at the estimate stage.

The inclusion of risk and inflation incorporates changes that may or may not take place in the future. This means that the description of the final cost as a cost limit may give the client a false sense of security.

Value Added Tax

As set out in NRM, p. 45, item 2.17, the NRM suggests that Value Added Tax (VAT) is excluded from the estimate on the basis of complexity. However, it is a significant area of expenditure and a comprehensive estimate of costs is not complete without a VAT assessment. If the intention is to provide a full picture to the client then VAT should be included. Most quantity surveying practices have sufficient expertise to provide this information and it is another source of fees.

5.2 METHOD OF MEASUREMENT

The estimate is usually measured as follows:

- Each group element and element of the building. For example, in NRM, p. 28, the group element of the superstructure comprises the elements frame, upper floors, roof etc.
- The measurement of each element using a particular unit of measurement. For example in NRM, p. 28, the frame is measured using square metres.
- The accuracy of the measurement is provided by using the standard method of measurement set out in the measurement rules described on p. 28. For example the gross internal floor area (GIFA) is used for measuring the upper floors.

The method of measurement is a new and separate standard method of measurement and comprises only superficial items or enumerated items. Items which might have been measured in cubic metres (eg excavation) are now measured in square metres. This means that measurements that do not conform to the NRM need to be converted. An example follows.

Substructure

The measurement is superficial in square metres. The rules clearly indicate that the ground or lowest floor slab is part of the substructure (NRM, p. 27) and is measured in square metres, not cubic metres (Table 5.4).

Table 5.4 Substructure method of measurement for basements.

Group element	Element	Unit	Measurement rules
1 Substructure		m^2	1 The area measured is the area of the lowest floor measured to the internal face of the external perimeter walls. 2 The area of the lowest floor shall be measured in accordance with the rules of measurement for ascertaining the gross internal floor area (GIFA). 3 Areas of basement to be shown separately. 4 The area of basement shall be measured in accordance with the rules of measurement for ascertaining the GIFA.

Source: RICS (2012, p. 27). Reproduced by permission of the RICS.

Compound items

Basements are measured and priced mainly in cubic metres and then converted to square metres by dividing the total cost by the GIFA area of the basement slab. This creates a compound item that enables accurate measurement and pricing and compliance with the requirements of the NRM (Table 5.5).

Table 5.5 Compound item: conversion of cubic to superficial measurement.

Piling		**Subcontractor quote**	**75,000**	
Excavate basement	2,000 m^3	× £2.50/m^3	5,000	
Earthwork support	400 m^2	× £5.00/m^2	2,000	
Remove surplus spoil	2,000 m^3	× £5.00/m^3	10,000	
Dewatering		Say	1,000	
Backfill	500 m^3	× £30.00/m^3	15,000	
Compact surface	500 m^2	× £2.00/m^2	1,000	
Waterproof membrane	500 m^2	× £2.00/m^2	1,000	
Basement slab	100 m^3	× £100.00/m^3	10,000	
Retaining walls	90 m^3	× £100.00/m^3	9,000	
Formwork	1,000 m^3	× £30.00/m^3	30,000	
Reinforcement	9 t	2,000/t	18,000	
Waterbar	300 m	× £10/m	3,000	180,000
GIFA floor area for basement				÷ 600 m^2
Rate per square metre				= £300/m^2
NRM Method of Measurement for Estimates		**600 m^2**	**× £300/m^2**	**= £180,000**

Frame

The frame (NRM, p. 28) of a building is usually reinforced concrete or steel. The measurement is superficial in square metres (Table 5.6). The use of volume for concrete or weight for steel is not used in this method. A compound item is also required to convert the normal measurement of cubic metres for reinforced concrete, or tonnes for steelwork, to a measurement in square metres.

Table 5.6 Superstructure estimates.

Group element	Element	Unit	Measurement rules
2 Superstructure	1 Frame	m^2	1 The area measured is the area of the floors related to the frame. 2 The area of the lowest floor shall be measured in accordance with the rules of measurement for ascertaining the gross internal floor area (GIFA).

Source: RICS (2012, p. 28). Reproduced by permission of the RICS.

Upper floors

Upper floors (NRM, p. 28) are measured superficially in square metres (Table 5.7). The rules usefully describe the inclusion of balconies and the like in this element and that they are to be measured separately. This is a good example of standardisation and consistency.

Table 5.7 Upper floors estimates.

Group element	Element	Unit	Measurement rules
2 Superstructure	2 Upper floors	m^2	1 The area measures is the total area of the upper floor(s). 2 The area of the lowest floor shall be measured in accordance with the rules of measurement for ascertaining the gross internal floor area (GIFA). 3 Sloping surfaces such as galleries, tiered terraces and the like are to measured flat on plan. 5 Areas for balconies, galleries, tiered terraces, service floors, walkways, internal bridges, external links, and roof to internal buildings shall be shown separately.

Source: RICS (2012, p. 28). Reproduced by permission of the RICS.

Space heating and air conditioning

Space heating and air conditioning (NRM 1, p. 32) is measured superficially in square metres (Table 5.8). The use of volume for space heating, air conditioning and ventilation systems for this element is not used in this method. A compound item is also required to convert the normal measurement of cubic metres for volume to a measurement in square metres.

Table 5.8 Space heating estimates.

Group element	Element	Unit	Measurement rules
5 Services	6 Space heating and air conditioning	m^2	1 The area measures is the area serviced by the system. 2 The area of the lowest floor shall be measured in accordance with the rules of measurement for ascertaining the gross internal floor area (GIFA). 3 Where more than one system is employed, the area measured for each system is the area serviced by the system. Areas to be measured using the rules of measurement for ascertaining the GIFA.

Source: RICS (2012, p. 32). Reproduced by permission of the RICS.

5.3 PRACTICAL APPLICATION: ESTIMATE LONDON ROAD BASEMENT

To illustrate the new measurement rules of the NRM for estimates the practical example (Table 5.9) provides a current interpretation. It comprises the measure for the basement for the London Road building. The measurement comprises the normal method of measuring excavation by cubic metres, formwork by square metres and reinforcement by the tonne. This is then converted to a compound item (Table 5.10) to follow the NRM methodology.

Table 5.9 London Road estimate, basement floor areas.

			Basement Floor Areas					
	48.00 ✓		Conference		*The figured dimensions are limited.*			
	12.00	576.00 ✓	4800 x 3 = 14400 ✓	595m^2				
1/2/	19.20 ✓		[Conference 4800 ✓					
Scaled]	2.00	19.20 ✓	19200 ✓		*The rest have to be scaled.*			
		595.20 ✓						
					A query sheet of problems should be provided to the architect and engineer.			
			To provide an accurate estimate for the basement the following items need to be measured and included in the estimate: Piling Excavate Remove surplus spoil Earthwork support Compact surfaces Retaining walls Basement slab Column bases Columns Ground beams Footings for staircases Column bases for steelwork Formwork Reinforcement Waterbar/Tanking This is shown at the end of the measurement.					

Table 5.9 London Road estimate, basement floor areas (*Continued*)

Basement Measurements
1: Substructure
1.2: Basement excavation

			Piling Quotation	Item	*Piling can be measured if information is available.*
	48.00 12.00 3.50	2,016.00	Basement excavation & Remove surplus spoil	2,083 m^3	*Structural floor heights have been scaled.*
Scaled]	19.20 2.00 3.50	67.20 2,083.20		2,083 m^3	*Measurements are taken in the normal way to enable the work to be measured and priced.*
2/	48.00 3.50	336.00	Earthwork support	423 m^2	*The adjustment to provide the measure in accordance with the NRM is shown at the end.*
2/	12.00 3.50	84.00			
2/	.40 3.50	2.80 422.80			
			$x^2 = 4800^2 + 2000^2$ $x^2 = 23.04 + 4000$ $x = \sqrt{27.04}$ $x = 5200$ (4800) 400		*This calculation is the additional length of the hypotenuse for the conference area.*
			Dewatering	Item	

Table 5.9 London Road estimate, basement floor areas (*Continued*)

Timesing	Dimensions	Squaring	Description	Quantity
			Basement Measurement	
	48.00		Compact surface	595 m²
	12.00	576.00		
½/	19.20		&	
	2.00	19.20	Waterproof membrane	595 m²
		595.20		
	48.00		Basement slab	179 m³
	12.00			
	.30	172.80		
½/	19.20			
	2.00			
	.30	5.76		
		178.56		
2/	48.00		Retaining walls	84 m³
	.20			
	3.50	67.20		
2/	12.00			
	.20			
	3.50	16.80		
2/	.40			
	.20			
	3.50	.41		
		84.41		

Timesing	Dimensions	Squaring	Description	Quantity			
			Basement Measurement				
2/2/	48.00		Formwork	846 m²			
	3.50	672.00	(both sides measured)				
2/2/	12.00						
	3.50	168.00					
2/2/	.40						
	3.50	5.60					
		845.60					
			Reinforcement to slab				
			178.56 m³ × 2% = 3.57				
			Reinforcement to walls				
			84.56 m³ × 2% = 1.69				
			5.26t	5.26 t			
2/2/	48.00	192.00	Waterbar/tanking	241 m			
2/2/	12.00	48.00					
2/	.40	.80					
		240.80					
2/	48.00		Backfill	423 m³			
	3.50		&				
	1.00	336.00	Working space				
2/	12.00						
	3.50						
	1.00	84.00					
2/	.40						
	3.50						
	1.00	2.80					
		422.80					

Table 5.10 Compound item for estimate of substructure basement.

Substructure (NRM, p. 27, items 1.1.3 and 1.1.4)				
Description	**Quantity**	**Rate**	**£**	**£**
Piling		Quotation	75,000	
Attendances on piling		10%	7,500	
Excavation	2,083 m^3	× £3/m^3	6,249	
Remove surplus spoil	2,083 m^3	× £5/m^3	10,415	
Earthwork support	423 m^2	× £5/m^3	2,115	
Dewatering		Item	5,000	
Compact surface	595 m^2	× £2/m^3	1,190	
Waterproof membrane	595 m^2	× £2/m^3	1,190	
RC slab	179 m^3	× £100/m^3	17,900	
RC retaining wall	85 m^3	× £125/m^3	10,500	
Formwork	846 m^2	× £30/m^2	25,380	
Reinforcement	5 t	× £2,000/t	10,000	
Working space	423 m^3	× £10/m^3	4,230	
Backfill	423 m^3	× £30/m^3	12,690	
Waterbar	241m	× £15/m	3,615	
GIFA floor area for basement			193,099	÷ 595 m^2
Rate per square metre				= £325/m^2
NRM Method of Measurement of Estimate for basement	**595 m^2**		**× £325/m^2**	**= £193,375**

Measuring and pricing the individual items of work enables an accurate quantity to be provided. The precise price per square metre can then be calculated. The same quantities also provide the basis for the measure of the cost planning elements of work.

5.4 SELF-ASSESSMENT EXERCISE: ESTIMATE LONDON ROAD RC FRAME

Our next exercise is to measure the quantities for the frame estimate.

Prepare a query sheet and compare the specification provided with the information schedules in the NRM.

Compare your own work with the proposed solution included in Appendix 7 (Appendix 7 is on the website (http://www.wiley.com/go/ostrowski/estimating)).

Self-assess your work on the assessment sheet included in Appendix 4 (Appendix 4 is on the website (http://www.wiley.com/go/ostrowski/estimating)).

To provide further assistance there are also dedicated websites at http://ostrowskiquantities.com and at Wiley Blackwell (http://www.wiley.com/go/ostrowski/estimating). It is hoped that the provision of this will go some way towards explaining the concepts and principles more clearly than using the printed word alone.

6 NRM 1 Cost Plans

6.1 PRACTICE AND PROCEDURE

Introduction

In this chapter we examine how to prepare cost plans using NRM 1 which includes a new and different standard method of measurement from that used for estimates. The procedures for building up the constituents are elaborated, which leads to a series of progressive cost plans. The elements are expanded to five levels of progressively greater detail. The examples indicate that the detail can be expanded to include workmanship, where this information is available. The specific method of measurement for cost plans is examined in detail and examples of different elements are provided. There are often several different units of measurement in each element of a cost plan. The referencing of these elements becomes elaborate. There are three major stages of cost plans which corresponds the development of the design. Each is described in detail. The practical example allows the procedure to be followed for the measurement of a basement cost plan and the self-assessment exercise allows practice to be undertaken for the measurement of the frame element of a cost plan.

Estimating and Cost Planning Using the New Rules of Measurement, First Edition. Sean D.C. Ostrowski.

Stages

The order for cost plans corresponds to RIBA Stage C Concept to Stage E Technical Design and to part of the Office of Government Commerce (OGC) Gateways Stage 2 Delivery Strategy and Stage 3A Design Brief and Concept Approval and part of Stage 3B Detailed Design Approval. This is set out in Table 6.1.

Table 6.1 Work stages for cost plans.

RIBA Work Stages		RICS formal cost estimating and elemental cost planning stages	OGC Gateways (Applicable to building projects)
			2 Delivery Strategy
Design	C Concept	Formal Cost Plan 1	3A Design Brief and Concept Approval
	D Design Development	Formal Cost Plan 2	
	E Technical Design	Formal Cost Plan 3	3B Detail Design Approval

Source: RICS (2012, p. 8). Reproduced by permission of the RICS.

The intent is to correlate cost plans with the RIBA Work Stages and the OGC Gateways. The three processes shown above concern the different elements of design, finance and construction. This means that the correlation cannot be exact and so there is some overlap of the stages. The function of the cost plan is to provide the financial information which enables the design developments to be related to the cost limit set out in the Estimate and to anticipate the tender with a high level of accuracy.

There are three stages of progressive cost plans. They are set out in NRM 1, pp. 54–5. The first cost plan (CP1) uses a condensed list of two elements which is restricted to two levels and a new method of measurement. The second (CP2) is based on the completion of the design development. It is '. . . *developed by cost checking of cost significant cost targets for elements* . . .'. It uses an expanded list of elements using three to five element levels. New measurement is only required for changes to CP1. The third (CP3) is based on '. . . *technical designs, specifications and detailed information for construction* . . .'. It is a progression of CP2 and is also to be used as the basis for appraising tenders. These stages are examined in detail later in this chapter.

The later stages of construction, stages F–L of the RIBA work stages, take the work to the post-practical completion stage. The NRM indicates that estimates should also be prepared in this period. At RIBA Stage F Production Information, a pre-tender estimate is to be prepared and at RIBA Stage H tender action, a post-tender estimate is to be prepared. The preparation of bills of quantities and work packages are the major constituents for pricing the anticipated costs at these stages of pre-tender and post-tender estimates. These quantities are measured on a trade basis and are beyond the scope of this textbook.

The overlap between Stages E, F, G and H (OGC Stages 3A, B and C) usually means that the final estimate, that anticipates the tender price, is only available when the tender documentation has been dispatched. These stages are set out in Table 6.2.

Table 6.2 Work stages for post-contract work.

RIBA Work Stages		RICS formal cost estimating and elemental cost planning stages	OGC Gateways (applicable to building projects)
Pre-Construction	F Production Information	Pre-tender estimate	
	G Tender Information		
	H Tender Action	Post-tender estimate	3C Investment Decision
Construction	J Mobilisation		
	K Construction to Practical Completion		4 Readiness for Service
Use	L Post Practical Completion		5 Operations Review and Benefits Realisation

6.2 ELEMENTS

Cost plans use an expanded schedule of elements. There are five levels of elements. Appendix E sets out the first three levels of the elements (Table 6.3).

Table 6.3 Elements for cost plans.

LEVEL 1 Group element	LEVEL 2 Element	LEVEL 3 Sub-element
3 Internal finishes	1 Wall finishes	1 Finishes to walls **Definition:** Applied finishes to internal wall surfaces, including specialist wall finishes for sports, public amenities and the like.
etc.	etc.	etc.

Source: RICS (2012, p. 348). Reproduced by permission of the RICS.

The three levels of Appendix E are then expanded to a total of five levels as Part 4, NRM, p. 143 as listed below and in Table 6.4.

Level 1 Group element. The primary group
eg Group element 3 Internal finishes.
Level 2 Elements. The elements that comprise the group element
eg 3.1 Wall finishes.
Level 3 Sub-elements. The parts that comprise the element
eg 3.1.1 Applied finishes
Level 4 Components. The measured parts of a sub-element
eg 3.1.1 plastering, painting
3.1.2 picture rails
Level 5 Sub-components. Other cost-significant items. Level 5 can only be used if sufficient information is provided concerning the sub-components. They can also provide supplementary information concerning workmanship, eg decorative features, bonding in brickwork, mortar specification etc.

Table 6.4 Five levels of elements for cost plans for wall finishes.

LEVELS	CONSTITUENT	Ref.	PAGE	UNIT	DESCRIPTION
1	Group element	3	143–348		Internal finishes
2	Element	1.1		m^2	Wall finishes
3	Sub-element	3.1.1		m^2	Finishes to walls
4	Component	3.1.1.1	143	m^2	Walls and columns: details to be stated
5	Sub-component	3.1.1.1.C1		m^2/lm/nr	Linear metres or as enumerated

Table 6.5 Five levels of measurement in cost plans for internal finishes.

LEVEL 1 **Group element 3: Internal finishes**

Group element 3 comprises the following elements:

3.1 Wall finishes
3.2 Floor finishes
3.3 Ceiling finishes

LEVEL 2 **Element 3.1: Wall finishes**

LEVEL 3 Sub-element	**LEVEL 4** Component	Unit	**LEVEL 5** Measurement rules for components	Included	Excluded
1 Finishes to walls **Definitions:** Applied finishes to internal wall surfaces, including specialist wall finishes for sports, public amenities and the like	1 Finishes to walls and columns: details to be stated 2 Picture rails, dado rails and the like: details to be stated 3 Proprietary impact and bumper guards, protection strips, corner protectors and the like: details to be stated	m^2 m nr/m	C1 Where components are to be enumerated, the number of components is to be stated C2 etc.	1 In-situ coatings applied to walls (eg plaster, render and roughcast) etc.	1 Fire protective coatingand paint sysyems to structural steel frames (included in element 2.1.1: Steel frames) etc.
etc.	etc.	etc.	etc.	etc.	etc.

Source: RICS (2012, p. 143). Reproduced by permission of the RICS.

6.3 METHOD OF MEASUREMENT FOR COST PLANS

The method of measurement for cost plans (Table 6.6) is a new standard method of measurement. It is only for use on cost plans. The method of measurement for cost plans is the tabulated rules of measurement for elemental cost plans as set out in Part 4 (NRM, p. 74). Some examples are described below in order to illustrate the more extensive measurement which is required.

Substructure, frame, upper floors, floor finishes and air conditioning

Table 6.6 Methods of measurement for cost plans for piling and basement.

	References		Description	Unit	Notes
Piling (NRM, p. 91)	**1.1.2**	1.1.2.1	Mats	m^2	
		1.1.2.2	Plant	Item	
		1.1.2.3	Transport	nr	
		1.1.2.4	Piles	nr	
		1.1.2.7	Disposal	m^3	
		1.1.2.9	Cutting tops	nr	
		1.1.2.10	Tests	Item	
		1.1.2.12	Pile caps	m	
Basement					
Lowest floor construction (NRM, p. 95)	**1.1.3**	1.1.3.1	Basement slab	m^2	
Basement excavation (NRM, p. 98)	**1.1.4**	1.1.4.1	Excavation	m^3	
		1.1.4.2	Remove surplus spoil		
		1.1.4.4	Earthwork support	m^2	
		1.1.4.1	Compact surface	Included	Included in ref. 7
		1.1.4.5	Waterproof membrane	m^2	
Basement retaining walls (NRM, p. 99)	**1.1.5**	1.1.5.1	Retaining walls	m^2	Surface area
		1.1.5.1	Backfill	Included	Included in ref. 1
		1.1.5.1	Formwork	Included	Included in ref. 1
		1.1.5.1	Reinforcement	Included	Included in ref. 1
		1.3.1.1.C5	Waterbar	m	
Specialist groundworks (NRM, p. 82)	**0.4.1**	0.4.1.1	Dewatering	Item	Specialist groundworks

Substructure (NRM, p. 89)

The measurement for basements (Table 6.7) includes both cubic and superficial quantities. The rules clearly indicate that excavation, and the disposal of excavated material is to be measured separately in cubic metres and earthwork support is to be measured in square metres. Although in theestimates all the items had to be measured and then combined into a single compound measurement of the square metres of the floor area, in cost plans each of the items is separate. Supplementary information concerning the rules and what is included and excluded provide a comprehensive template of information. The practical application in Section 6.5 provides an example of how this is executed.

Table 6.7 Basement.

Element 1.1: Substructure, Sub-element 1.1.4, Basement excavation

Sub-element	Component	Unit	Measurement rules for components	Included	Excluded
1 Basement excavation **Definitions:** Bulk excavation required for construction of floors below ground level.	1 Basement excavation: including average depth of excavation, to be stated. 2 Disposal of excavated material: details to be stated. 3 Extra for disposal of contaminated excavated material: details to be stated. 4 Earthwork support: details to be stated. 5. Additional excavation: details to be stated.	m^3 m^2	C1 Where components are to be enumerated, the number of components is to be stated. C2 etc.	1 Bulk excavation to form basement and the like. etc.	1 Excavation and earthworks to forming new site contours and adjust existing site levels (included in sub-element 8.1.2, Preparatory groundworks). etc.
etc.	etc.	etc.	etc.	etc.	etc.

Source: RICS (2012, p. 98). Reproduced by permission of the RICS.

Frame (NRM, p. 105)

The measurement for a reinforced concrete frame (Table 6.8) is in linear metres for both columns and beams which are to be measured separately. Supplementary information concerning the rules and what is included and excluded, provides a comprehensive template of information. This has been briefly examined in Chapter 4 and is elaborated here as it is a significant part of measurement for cost plans.

Item 2.1.4.4 (NRM, p. 106) describes *'Extra over walls for forming openings in walls . . .'* as a measureable item that is to be enumerated. Item 13 in the 'Included' column states that *'Forming openings for doors, windows, screens and the like'* is included. The meaning is that openings are to be measured as separate items and included in the measure.

Item 2.1.4.5 (NRM, p. 106) describes *'Designed joints . . .'* as a measureable item in linear metres. Item 9 in the 'Included' column states that *'Designed joints'* are included. The meaning is that designed joints are to be measured as separate items and included in the measure.

Table 6.8 Reinforced concrete frames.

Sub-element	Component	Unit	Measurement rules for components	Included	Excluded
4 Concrete frames **Definitions** Concrete columns and beams.	1 Columns details, including number (nr) of columns, column sizes, concrete grade, reinforcement rate (kg/m^3) and type of formwork finish, details to be stated. 2 Beams details, including number (nr) of columns, column sizes, concrete grade, reinforcement rate (kg/m^3) and type of formwork finish, to be stated.	m	C1 Columns. The linear length measured is is the distance between the top of the slab, pile cap or ground beam (as appropriate) and the soffit of the beam atttached to the next floor level (or to the soffit of the suspended slab if no beams). C2 etc.	1 Beams. 2 Columns, blade columns and the like. etc.	1 Upper floors (included in element 2.2 Upper floors). etc.
etc.	etc.	etc.	etc.	etc.	etc.

Source: RICS (2012, p. 105). Reproduced by permission of the RICS.

Upper floors (NRM, p. 109)

Upper floors (Table 6.9) are measured superficially in square metres. In addition, formwork to the edges and designed joints are measured in linear metres and surface treatments are measured in square metres. Supplementary information concerning the rules and what is included and excluded, provides a comprehensive template of information.

Table 6.9 Upper floors in reinforced concrete.

Element 2.2: Upper floors

Sub-element	Component	Unit	Measurement rules for components	Included	Excluded
1 Concrete floors **Definitions:** Reinforced and post-tensioned concrete suspended floors.	1 Suspended floor slabs: details, including thickness (mm); concrete strength (N/mm^2) and type of formwork finish, details to be stated. 2 Edge formwork: details of formwork finish to be stated.	m^2 m	C1 The area measured is the area of the upper floors. The area is measured using the rules of measurement for ascertaining the gross internal floor area (GIFA). No deduction is to be made for beams which form an integral part of the upper floor C2 etc.	1 Concrete suspended flooring, including: - podium slabs form roofs to basements - transfer structures - balconies (internal and external) which are an integral part of the suspended floor construction - mezzanine floors - service floors and the like - galleries, tiered terraces and the like - walkways, internal bridges and the like External corridors/bridges forming links between buildings, including supporting frames - beams which form an integral part of the floor in framed buildings - floor beams in unframed buildings - roofs to internal buildings, where an integral part of the upper floor construction. etc.	1 Basement roofs ie where not acting as a podium slab or transfer slab (included in element 2.3: Roof). etc.
etc.	etc.	etc.	etc.	etc.	etc.

Source: RICS (2012, p. 10). Reproduced by permission of the RICS.

Floor finishes (NRM, p. 144)

Floor finishes (Table 6.10) includes extensive series of measurements and is no longer restricted to a simple GIFA multiplied by a compound rate. This includes linear items such as skirtings and enumerated items such as mat wells. Supplementary information concerning the rules and what is included and excluded, provides a comprehensive template of information.

Table 6.10 Floor finishes.

Element 3.2: Floor finishes

Sub-element	Component	Unit	Measurement rules for components	Included	Excluded
1 Finishes to floors **Definitions:** Applied finishes to floor surfaces, including specialist floors to sports facilities, public amenities and the like.	1 Finishes to floors: details to be stated. 2 Specialist flooring systems: details to be stated. 3 Skirtings and the like: details to be stated. 4 Mat wells and mats: details to be stated. 5 Finishes to swimming pool tanks, including tank linings: details to be stated.	m^2 m nr m^2	C1 Where components are to be enumerated, the number of components is to be stated. C2 The length of linear components measured is their extreme length, over all obstructions. C3 The area measured for each type of floor finish is the surface area of the floor to which the surface is applied. C4 etc.	1 Non-structural screeds, including underscreed damp-proof membranes. 2 Latex screed (ie levelling screed) 3 Chemical surface hardeners and sealers applied to screed. 4 etc.	1 Fire protective coatings and paint systems to structural steel frames (included in sub-element 2.1.1 Steel frames). 2 Structural screeds(including sub-element 1.4.1: Ground floor slabbed and suspended floor construction or element 2.2 Upper floors, as appropriate). 3 etc.
etc.	etc.	etc.	etc.	etc.	etc.

Source: RICS (2012, p. 144). Reproduced by permission of the RICS.

Air conditioning (NRM, p. 183)

Air conditioning (Table 6.11) is measured superficially in square metres. Testing and commissioning is added as a percentage to each sub-element rather than as a separate item. Supplementary information concerning the rules and what is included and excluded, provides a comprehensive template of information.

Table 6.11 Air conditioning.

Sub-element	Component	Unit	Measurement rules for components	Included	Excluded
7 Central air conditioning **Definitions:** Systems where air treatment is performed at a central point and air is distributed to the spaces and locations being treated.	1 Central air conditioning systems, details to be stated. 2 Testing of installation. 3 Commissioning of installation.	m^2 %	C1 The area measured is the area serviced by the system (ie the area of the rooms and circulation spaces that are served by the system, which is not necessarily the total gross internal floor area (GIFA) of the building. The area served is measured using the rules of measurement for ascertaining the internal floor area (GIFA).	1 Plenum air heating systems. 2 VAV (variable air volume) and constant volume air conditioning systems. 3 Dual duct and induction air conditioning systems.	1 Heat sources included in element 5.5 Heat source. 2 Local cooling and air treatment eg local comfort cooling, included in sub-element 5.6.8 Local air conditioning.
etc.	etc.	etc.	etc.	etc.	etc.

Source: RICS (2012, p. 183). Reproduced by permission of the RICS.

6.4 COST PLANS

CP1 The tabulated rules are set out in NRM 1 pp. 67–73 and Appendix E (pp. 347–51). The first cost plan uses a condensed list of two levels of elements and a new method of measurement. Cost plan 1 is on p. 54 and is set out in Table 6.12.

Table 6.12 Cost Plan 1.

Formal Cost Plan 1	
(i)	This is the first formal cost plan. It coincides with the completion of the concept design at the point where the scope of the works is fully defined and key criteria are specified but no detailed design has commenced.
(ii)	Cost Plan 1 will provide the frame of reference for Cost Plan 2.
(iii)	The key information required from the employer and other project/design team members to enable preparation of the Formal Cost Plan 1 is set out in Appendix G of these rules.
(iv)	For Cost Plan 1, a condensed list of elements is used, which will be developed into a full list of elements, sub-elements and components as more design and other information becomes available as the building project progresses.
(v)	Quantities for building works shall be determined in accordance with *Part 4: Tabulated rules of measurement for elemental cost planning (ie group elements 1–9).*
(vi)	Where insufficient design information is available from which to quantify building works in accordance with the rules of measurement for elemental cost planning, then the quantity measured is to be the GIFA.
(vii)	It is likely that a number of alternative concept designs will be considered at this time.

Source: RICS (2012, p. 54). Reproduced by permission of the RICS.

CP2 This is based on the completion of the design development. It is '. . . *developed by cost checking of cost significant cost targets for elements* . . .' (NRM 1, p. 54). It uses an expanded list of elements using from three to five element levels. CP2 requires the completion of the design as Stage D Design Development. However, to suggest that the design is frozen at this point is rather premature. In addition, by implication, there is no new measurement unless there is some new design. Again, this is rather premature as the value engineering exercises are likely to continue. CP2 is at NRM 1, p. 54 and is set out in Table 6.13.

Table 6.13 Cost Plan 2.

Formal Cost Plan 2	
(i)	This is the second formal cost plan which coincides with the completion of the design development. Formal Cost Plan 2 is a progression of Formal Cost Plan 1. It is developed by cost checking cost significant cost targets for elements as more detailed information is made available from the design team.
(ii)	Cost Plan 2 will provide the frame of reference for Cost Plan 3.
(iii)	The key information required from the employer and other project/design team members to enable preparation of the Formal Cost Plan 2 is set out in Appendix G of these rules.
(iv)	The cost checks are to be carried out against each pre-established cost target.
(v)	Quantities for building works shall be determined in accordance with *Part 4: Tabulated rules of measurement for elemental cost planning (ie group elements 1–9).*
(vi)	Where insufficient design information is available from which to quantify building works in accordance with the rules of measurement for elemental cost planning, then the quantity measured is to be the GIFA.

Source: RICS (2012, p. 54). Reproduced by permission of the RICS.

CP3

This is based on '. . . *technical designs, specifications and detailed information for construction* . . .'. The NRM indicates that at this stage the construction drawings will be available. These are the detailed drawings at scale 1:50 and all the completed details, sections and specifications. CP3 is a progression from CP2 and is also to be used for appraising tenders. The full completion of the design at this stage is unlikely. CP3 is given in NRM, p. 55 and is set out in Table 6.14.

Table 6.14 Cost Plan 3.

Formal Cost Plan 3	
(i)	This third formal cost plan stage is based on technical designs, specifications and detailed information for construction. Formal Cost Plan 3 is a progression of Formal Cost Plan 2. It is developed by cost checking cost significant cost targets for elements as more detailed information is made available from the design team.
(ii)	Cost Plan 3 will provide the frame of reference for appraising tenders.
(iii)	The key information required from the employer and other project/design team members to enable preparation of the Formal Cost Plan 3 is set out in Appendix G of these rules.
(iv)	The cost checks are to be carried out against each pre-established cost target.
(v)	Quantities for building works shall be determined in accordance with *Part 4: Tabulated rules of measurement for elemental cost planning (ie group elements 1–9).*
(vi)	Where insufficient design information is available from which to quantify building works in accordance with the rules of measurement for elemental cost planning, then the quantity measured is to be the GIFA.

Source: RICS (2012, p. 55). Reproduced by permission of the RICS.

6.5 PRACTICAL APPLICATION: COST PLAN LONDON ROAD BASEMENT

To illustrate the new measurement rules of the NRM for cost plans the practical example below provides a current interpretation. It comprises the measure and pricing for the basement for the London Road office building. The measurement comprises the normal method of measuring concrete by cubic metres, formwork by square metres and reinforcement by the tonne. This is then converted to the NRM methodology. It should be noted that the items in italics are measured in the normal way but are not in accordance with the NRM methodology. These items are incorporated into those items that comply with the NRM.

An example of part of a cost plan for the basement is given in Table 6.15.

Table 6.15 London Road, priced substructure cost plan.

			Basement Floor Areas		
	48.00 ✓		Conference 4800 x 3 = 14400 ✓ 4800 ✓ 19200 ✓		*The figured dimensions are limited.*
	12.00 ✓	576.00 ✓			
½/	19.20 ✓		[Conference		*The rest have to be scaled.*
Scaled]	2.00 ✓	19.20 ✓			
		595.20 ✓			*A query sheet of problems should be provided to the architect.*

Table 6.15 London Road, priced substructure cost plan (*Continued*)

			Basement Cost Plan				
			Piling: Quotation [1.1.2.4	Item		75,000	00
			If the piling information is available it can be measured as Ref. 1.1.2.4 Piled Foundations NRM 1 p. 91				
			Mats [1.1.2.1	m^2			
			Plant [1.1.2.2.	Item			
			Transport [1.1.2.3	nr			
			Disposal [1.1.2.7	m^3	10%	7,500	00
			Cutting tops [1.1.2.9	nr			
			Tests [1.1.2.10	Item			
			Pile caps [1.1.2.12	m			

Table 6.15 London Road, priced substructure cost plan (*Continued*)

			Description	Quantity	Rate	Total	
			Basement Cost Plan				
	48.00		Basement excavation	2,083m³	£3	6,249	00
	12.00		[1.2.4.1				
	3.50	2,016.00	&				
½/	19.20		Remove surplus spoil	2,083m³	£5	10,415	00
	2.00		[1.2.4.2				
	3.50	67.20					
		2,083.20					
			References are the Basement excavation NRM 1 p. 98.				
			Pricing to include backfill				
2/	48.00		Earthwork support	423m²	£5	2,115	00
	3.50	336.00	[1.2.4.4				
2/	12.00		*References are the Basement excavation NRM 1 p. 98.*				
	3.50	84.00					
2/	.40						
	3.50	2.80					
		422.80					
			Dewatering	Item		5,000	00
			[0.4.1				
			References are the Specialist groundworks NRM 1 p. 82				

Table 6.15 London Road, priced substructure cost plan (*Continued*)

Timesing	Dimensions	Squaring	Description	Quantity	Rate	£	p
			Basement Cost Plan				
	48.00		*Compact surface*	~~595m^2~~	*Not included in NRM measure*		
	12.00	*576.00*	*[Included Ref 7*				
½/	*19.20*		*References are the Basement excavation NRM 1 p. 98*				
	2.00	*19.20*					
		595.20					
			&				
			Waterproof membrane [1.2.4.5	595m^2	£4	2,380	00
			Pricing for waterproof membrane to include compacting the surface of the excavation				
			Compact 595m^2 x £2/m^3 = £1,190				
			Membrane 595m^2 x £2/m^2 = £1,190				
			2,380				
			£2,380 ÷ 595 = £4/m^2				
			Basement slab				
			Measurement is in square metres. Pricing is in cubic metres. The cubic measure below for the concrete will need converting to a compound price for square metres as follows.	~~179m^3~~	*Not included in NRM measure*		
	48.00						
	12.00						
	.30	*172.80*					
½/	*19.20*		*RC 179m^3 x £100/m^3 = 17,900*				
	2.00		*Rebar 3.57t x £2,000/t = 7,140*		*References are the Ground floor construction NRM 1 p. 70*		
			£25,040				
	.30	*5.76*	*£25,040 ÷ 595 = £42/m^2*				
		178.56					
			Reinforcement to slab		*Not included in NRM measure*		
			[Included Ref 1.1.3.1 Note	~~3.57t~~			
			178.56m^3 x 2 % = 3.57t				
			Basement slab (Area as above) [1.1.3.1	595m^2	£42	24,990	00

Timesing	Dimensions	Squaring	Description	Quantity	Notes
			Basement Cost Plan		
			Retaining walls		
2/	*48.00*		*Retaining walls*	~~*85*m^3~~	*Not included in NRM measure*
	.20				
	3.50	*67.20*			
2/	*12.00*				
	.20				
	3.50	*16.80*			
2/	*.40*				
	.20				
	3.50	*.56*			
		84.56			
2/2/	*48.00*		*Formwork (both sides measured) [Included Ref 1.1.5.1.C1.1*	~~*846*m^2~~	*Not included in NRM measure*
	3.50	*672.00*			
2/2/	*12.00*		*References are the Basement retaining walls NRM Page 99*		
	3.50	*168.00*			
2/2/	*.40*		$x^2 = 4800^2 + 2000^2$		
	3.50	*5.60*	$x^2 = 23.04 + 4000$		
		845.60	$x = \sqrt{27.04}$		
			$x = 5200$		
			(4800)		
			400		
			Reinforcement to walls [Included Ref 1.1.5.1 Note		
			*84.56*m^3 *x 2% = 1.69t*	~~1.69 t~~	*Not included in NRM measure*
			References are the Basement retaining wall NRM 1 p.99		

Table 6.15 London Road, priced substructure cost plan (*Continued*)

2/	*48.00*		*Backfill*	~~*423m^3*~~	*Not included in NRM measure*		
	3.50						
	1.00	*336.00*	*[1.1.5.1.C1.1*				
2/	*12.00*		*&*				
	3.50		*Working space*				
	1.00	*84.00*	*[1.1.5.1.C1.1*				
2/	*.40*						
	3.50						
	1.00	*2.83*					
		422.83					
			Measurement for retaining walls is in square metres. Pricing for RC is in cubic metres. The cubic measure below for the concrete will need converting to a price for square metres as follows.				
			RC 85m^3 x £125/m^3 = 10,625				
			Rebar 1.69t x £2,000/t = 3,380				
			Fmwk 846m^2 x £30/m^2 = £25,380				
			Working Space & Backfill				
			423m^3 x £30/m^3 = £12,690				
			£ 52,075				
			£52,075 ÷ 423= £123/m^2				
			Calculations from above and previous page				
2/	48.00		Retaining walls	423m^2	£123	52,029	00
	3.50		[1.1.5.1.				
2/	12.00	336.00					
	3.50		*References are the Basement retaining wall NRM 1 p. 99*				
2/	.40	84.00					
	3.50	2.80					
		422.80					

Table 6.15 London Road, priced substructure cost plan (*Continued*)

			Basement Cost Plan				
2/2/	48.00	192.00	Waterbar	241m	£15/m	3,615	00
2/2/	12.00	48.00	[1.1.5.1.C5				
2/	.40	.80					
		240.80					

Table 6.15 London Road, priced substructure cost plan (*Continued*)

			Description			£	p
			Basement Cost Plan				
			Piling			75,000	00
						7,500	00
			Basement excavation			6,249	00
			Remove surplus spoil			10,415	00
			Earthwork support			2,115	00
			Dewatering			5,000	00
			Waterproof membrane			2,380	00
			Basement slab			24,990	00
			Retaining walls			52,029	00
			Waterbar			3,615	00
			NRM Method of Measurement of Cost Plan for Basement			**189,293**	**00**

6.6 SELF-ASSESSMENT EXERCISE: COST PLAN LONDON ROAD RC FRAME

Our next exercise is to prepare a cost plan for the reinforced concrete frame.

Prepare a query sheet and compare the specification provided with the information schedules in the NRM.

Compare your own work with the proposed solution included in Appendix 8 (Appendix 8 is on the website (http://www.wiley.com/go/ostrowski/estimating)).

Self-assess your work on the assessment sheet included in Appendix 4 (Appendix 4 is on the website (http://www.wiley.com/go/ostrowski/estimating)).

To provide further assistance there are dedicated websites at http://ostrowski quantities.com and at Wiley Blackwell (http://www.wiley.com/go/ostrowski/estimating). It is hoped that the provision of this will go some way towards explaining the concepts and principles more clearly than using the printed word alone.

7 Information

7.1 INTRODUCTION

The schedule of information required to prepare cost plans is a useful innovation. The informal arrangement of providing estimates based on whatever information is available has now been replaced with formal stages for information release which will then allow the progressive preparation of the estimate and cost plans. The schedules of information provide a clear indication of what information is to be made available at each stage. The level of information and the transparency of the process will be much improved. This discipline increases the coordination between the designers and will highlight to all parties the areas where information remains outstanding. The acceptance and implementation of these information schedules by the design team as stages in the design process that need to be completed, is one of the major challenges of the NRM 1.

Estimating and Cost Planning Using the New Rules of Measurement, First Edition. Sean D.C. Ostrowski.

7.2 INFORMATION REQUIREMENTS FOR ESTIMATES

The information requirements for estimates are set out in NRM, section 2.3, pp. 20–1. There is a separate schedule of information for the employer and each of the major designers and they will be examined in turn.

Information required from the employer

Table 7.1 indicates the way that the extent of the information that is available can be incorporated into the estimate. Several of the items are information that the client expects the consultants to provide rather than the client being able to provide them. For example the floor areas would be provided by the architect or quantity surveyor rather than the client. The availability column indicates the information normally available from the employer at this stage.

Table 7.1 Availability of information from the employer.

2.3.1	EMPLOYER'S INFORMATION	AVAILABILITY
a	Location and availability	✓
b	Building use	✓
c	Floor area and accommodation	
d	Requirements	
e	Brief: quality/sustainability and fit out	
f	Enabling works	
g	Programme	
h	Restraints	
i	Site conditions	
j	Budget/cash flow	
k	Procurement	
l	Life span	
m	Storey heights	
n	M&E requirements	
o	Fees/costs/inflation/VAT	
p	Other considerations	

Source: Derived from RICS (2012), p. 20. Reproduced by permission of the RICS.

Information required from the architect

Table 7.2 indicates the way that the extent of the information that is available can be incorporated into the estimate. Several of the items are information that the architect would not normally expect to provide at this stage. The availability column indicates the information normally available from the architect at this stage.

Table 7.2 Availability of information from the architect.

2.3.2	ARCHITECT'S INFORMATION	AVAILABILITY
a	Sketches	✓
	Building use	✓
	Floor plans	✓
	Roof plans	✓
	Elevations	✓
	Sections	
b	Schedule of areas	
c	Storey height	✓
d	Accommodation	
e	Car parking	✓
f	Specification	✓
g	Environmental	
h	Site constraints	
i	Planning constraints	
j	Fit out	
k	Risk	

Source: Derived from RICS (2012), p. 21. Reproduced by permission of the RICS.

Information required from the services engineer

The specification for heating/air conditioning is likely to be based on the volume of the building to be serviced. The electrical specification will be a range of options which will be dependent on how heavily serviced the building will be. The availability column (Table 7.3) indicates the information normally available from the services engineer at this stage.

Table 7.3 Availability of information from the services engineer.

2.3.3	M&E SERVICES INFORMATION	AVAILABILITY
a	Specification	✓
b	Sustainability	
c	Utilities/Service connections	✓
d	Risk	

Source: Derived from RICS (2012), p. 21. Reproduced by permission of the RICS.

Information required from the structural engineer

The provision of the structural layout is one of the most significant elements of the estimate. The availability column (Table 7.4) indicates the information normally available from the structural engineer at this stage.

Table 7.4 Availability of information from the structural engineer.

2.3.4	STRUCTURAL INFORMATION	AVAILABILITY
a	Ground conditions	✓
b	Specifications	✓
c	Risk	

Source: Derived from RICS (2012), p. 19. Reproduced by permission of the RICS.

These schedules apply after the site has been acquired. The initial approximate estimate that precedes this remains informal and will have little of this information. However, this first approximate estimate will be the report that enables the site to be purchased and the reported figure in this initial, preliminary and approximate estimate will be the most enduring. It is a normal part of the service provided by the quantity surveyor to furnish the client with an estimate when one is requested. The most difficult part of implementing this information schedule is resisting the demands of the client for an estimate until the information is available.

7.3 INFORMATION REQUIRED FOR THE COST PLANS

The information requirements for the cost plans are set out in NRM 1, p. 53 and Appendix F. Each cost plan has a separate schedule for the employer and for each of the major designers and each will be examined in turn.

Information required from the employer

Table 7.5 Information requirements from the employer for CP1.

CP1	EMPLOYER'S INFORMATION
i	Cost limit
ii	Brief and fit out
iii	Programme
iv	Procurement
	• Contract
	• Phasing
	• Facilitating works
	• Fees
	• Insurance
	• Sundry
	• Planning gains
	• Risk
	• Inflation
	• VAT
	• Allowances
v	Post Completion
vi	Authority to proceed to next stage

Source: RICS (2012), Summary of Appendix F, p. 352. Reproduced by permission of the RICS.

The value engineering process begins at this stage. There will be several alternative solutions showing changes to design and specification and to the financial consequences. The confirmation of the cost limit depends on the result of these value engineering exercises and will continue into CP2 and beyond. The fit-out requirements will only be available if the building has a specific function. Several elements of the fit out may be postponed to the stage of work after completion.

Several of the items are information that the client would normally expect the consultants to provide. The programming requirements, including a design schedule, start and completion dates, and critical dates require formal programming expertise. These are unlikely to be available nor are they included in the fee for the preparation of the estimate. The procurement stategy will develop through the progressive cost plans rather than be fixed at this early stage. A risk register is prepared by the consultants to provide the client with an assessment of the likely impact of the problems that may be encountered. This is possible, if the employer is a knowledgable client with experience in this particular type of development. The extent to which information can be reasonable expected at this stage is a problem and may prejudice the authority to proceed to the next design stage and CP2. However, by using these information schedules the client and the design team will be fully aware of the extent to which the requirements of CP1 have been adhered to. Table 7.6 indicates the extent of the information that is normal for CP1.

Table 7.6 Availability of information.

CP1	EMPLOYER'S INFORMATION	AVAILABILITY AT CP1
I	Cost limit	
ii	Brief and fit out	✓
iii	Programme	
iv	Procurement	✓
	• Contract	
	• Phasing	✓
	• Facilitating works	✓
	• Fees	
	• Insurance	✓
	• Sundry	
	• Planning gains	✓
	• Risk	
	• Inflation	✓
	• VAT	✓
	• Allowances	
v	Post Completion	
vi	Authority to proceed to next stage	

Table 7.7 sets out the information requirements from the employer for CP1, 2 and 3. This indicates that virtually all the information is required for CP1 and is then extensively repeated for CP2 and again for CP3.

Table 7.7 Employer's information for cost plans.

EMPLOYER'S INFORMATION	CP1	CP2		CP3	
Confirm previous CP			✓		✓
Confirm alternatives		i	✓	i	✓
i. Cost limit	✓	ii	✓	ii	✓
ii. Brief and fit out	✓	iii	✓	iii	
iii. Programme	✓	iv	✓	iv	
iv. Requirements		v		v	
• Procurement	✓		✓		✓
• Contract	✓		✓		✓
• Phasing	✓		✓		✓
• Facilitating works	✓		✓		✓
• Fees	✓		✓		✓
• Insurance	✓		✓		✓
• Sundry	✓		✓		✓
• Planning gains	✓		✓		✓
• Risk	✓		✓		✓
• Inflation	✓		✓		✓
• VAT	✓		✓		✓
• Allowances	✓		✓		✓
v. Post Completion	✓	vi		vi	✓
Accept previous other matter		vii	✓	vii	✓
vi. Authority to proceed to next stage	✓	viii	✓	viii	✓

Information required from the architect

Table 7.8 sets out the information requirements from the employer for CP1. This indicates that virtually all the information is required for CP1.

Table 7.8 Information requirements from the architect for CP1.

CP1	ARCHITECT'S INFORMATION
i	**Design**
	• GA plans
	• GA elevations
	• GA sections
	• Landscaping
	• Functions
	• Construction details
	• Interfaces
	• Room concepts
	• Constraints

(*Continued*)

Table 7.8 (*Continued*)

CP1	ARCHITECT'S INFORMATION
ii	**Floor areas**
iii	**Outline specification**
	• Elements
	• Quality
	• Components, materials
	• Acoustics
	• Outline performance criteria
	• Finishes
	• Alternative specifications
iv	**Room data sheets**
v	**Fittings schedule**
vi	**Strategies**
	• Sustainability
	◦ BREEAM
	◦ Building Regulations
	◦ Sustainability
	◦ Renewable energy
	◦ Specific requirements
	• Car parking
	• Vertical movement
	• IT
	• Fire
	• Acoustics
	• Security
	• DDA
	• Window cleaning
	• Refuse
	• Public art
	• Conservation
	• Sundry
vii	**Reports**
	• Archaeological
	• Topographical
viii	**Construction methodology**
ix	**Fit out**
x	**Risk**

Source: RICS (2012), Summary of NRM, Appendix F, p. 353. Reproduced by permission of the RICS.

The extent to which it is reasonable to expect all the information shown in Table 7.8 to be available for CP1 is problematical. Several of the items are information that the architect expects to be provided by other consultants to the project. Table 7.9 indicates the extent of information that would normally be available at CP1.

Table 7.9 Availability of information.

CP1	ARCHITECT'S INFORMATION	AVAILABILITY AT CP1
i	**Design**	
	• GA plans	✓
	• GA elevations	✓
	• GA sections	✓
	• Landscaping	
	• Functions	
	• Construction details	
	• Interfaces	
	• Room concepts	
	• Constraints	✓
ii	**Floor areas**	✓
iii	**Outline specification**	
	• Elements	✓
	• Quality	✓
	• Components, materials	
	• Acoustics	
	• Outline performance criteria	✓
	• Finishes	
	• Alternative specifications	
iv	**Room data sheets**	
v	**Fittings schedule**	

(Continued)

Table 7.9 (*Continued*)

CP1	ARCHITECT'S INFORMATION	AVAILABILITY AT CP1
vi	**Strategies**	
	• Sustainability	
	◦ BREEAM	
	◦ Building Regulations	✓
	◦ Sustainability	
	◦ Renewable energy	
	◦ Specific requirements	
	• Car parking	✓
	• Vertical movement	
	• IT	
	• Fire	
	• Acoustics	
	• Security	
	• DDA	
	• Window cleaning	
	• Refuse	
	• Public art	
	• Conservation	
	• Sundry	
vii	**Reports**	
	• Archaeological	
	• Topographical	
viii	**Construction methodology**	
ix	**Fit out**	
x	**Risk**	

Several items are likely to have a minimum amount of information at this stage, for instance:

- Landscaping
- Construction details
- Key details
- Detailed floor schedules (an external consultant may have been appointed)
- Acoustics
- Finishes schedules
- Alternative specifications (for CP2)
- Room data sheets (for CP2/3)
- Fittings
- Sustainability
- IT
- Fire
- Security
- Window cleaning
- Public art
- Phasing
- Fit out
- Risk

The information that is required from the architect at CP1 is extensively repeated for CP2 and again for CP3. However, the information requirements can be progressively populated from CP1 to CP3 as will be described in Section 7.4.

Information required from the services engineer

Table 7.10 indicates the NRM requirements for information available from the services engineer and the extent of the information that would normally be available at CP1.

The information that is required from the service engineer for CP1 is extensively repeated for CP2 and again for CP3. However, the information requirements can be progressively populated from CP1 to CP3 as described in Section 7.4.

Table 7.10 Information required from the services engineer.

	M&E SERVICES INFORMATION	AVAILABLE AT CP1
i	**Drawings**	
	• GA	✓
	• Schematic	✓
	• Plant room	
	• Diagrams	
	• Layouts	

(Continued)

Table 7.10 (*Continued*)

	M&E SERVICES INFORMATION	AVAILABLE AT CP1
ii	**Specification**	
	• Mechanical	
	• Electrical	
	• Transport	
	• Protective	
	• Communications	
	• Security	
	• BMS	✓
	• Special	
	• Plant schedule	✓
	• Duties	
	• BWIC	✓
	• Alternatives	
iii	**Strategies**	
	• Environmental	
	• BREEAM	
	• Building Regulations	✓
	• Sustainability	
	• Renewable energy	
	• Specific requirements	
	• Vertical movement	✓
	• Removals	
	• Commissioning	
iv	**Reports**	
	• Existing services	✓
	• Abnormals	
v	**Utilities**	
	• Connections	
	• Upgrading	✓
	• Diversions	✓
	• Quotations	
vi	**Facilitating works**	
vii	**Risk**	

Information required from the structural engineer

Table 7.11 indicates the NRM requirements for information available from the structural engineer and the extent of the information that would normally be available at CP1.

Table 7.11 Information required from the structural engineer.

	STRUCTURAL INFORMATION	AVAILABLE AT CP1
i	**Reports: Desktop**	
	• Contamination	✓
	• Geotechnical	✓
	• Bombs	✓
ii	**Reports: Fieldwork**	
	• Contamination	
	• Geotechnical	
iii	**Environmental risk**	
iv	**Ground conditions**	✓
v	**Design**	
	• GA	✓
	• Frame	✓
	• Layout	
	• Sections	
	• Foundations	✓
	• Substructure sections	✓
	• Drainage	✓
vi	**Levels**	✓
vii	**Specification**	
	Elements	
	• Components	
	• Materials	
	• Loadings	
	• Piling	✓
	• BWIC services	
	• Steelwork	
viii	**Reinforcement**	
ix	**Steelwork**	✓
x	**Methodologies**	
	• Facilitating works	✓
	• Temporary works	
	• Alterations	
	• Drainage	
xi	**Risk**	

The information that is required from the structural engineer for CP1 is extensively repeated for CP2 and again for CP3. However, the information requirements can be progressively populated from CP1 to CP3 as described in Section 7.3.

7.4 PROGRESSIVE PROVISION OF INFORMATION

It is now possible to prepare information requirements can be progressively populated from the estimates to each of the cost plans. Although each item needs measuring and pricing the initial estimates and cost plans can provide outline information and indicative prices. As the information becomes available in detail so the later cost plans can provide more detailed prices. This allows the design team to prepare each element in turn rather than try to complete all the elements simultaneously. Information from specialist consultants and subcontractors can be made available for the final cost plan. Examples of progressive schedules of information are set out below.

Progressive employer's information requirements

Table 7.12 Progressive employer's information requirements.

EMPLOYER'S INFORMATION	Estimate	CP1	CP2	CP3
Location and availability	✓			
Building use	✓			
Accommodation	✓			
Requirements	✓			
Enabling works		✓		
Restraints	✓			
Site conditions	✓			
Budget/cash flow			✓	
Life span				✓
M&E requirements				✓
Confirm previous CP			✓	✓
Confirm alternatives			✓	✓
Cost limit		✓	✓	✓

Table 7.12 (*Continued*)

EMPLOYER'S INFORMATION	Estimate	CP1	CP2	CP3
Brief: quality/sustainability		✓		
Fit out				✓
Programme				✓
Requirements				
• Procurement				✓
• Contract				✓
• Phasing			✓	
• Facilitating works		✓		
• Fees			✓	
• Insurance			✓	
• Sundry				✓
• Planning gains		✓		
• Risk				✓
• Inflation			✓	
• VAT			✓	
• Allowances				✓
Post Completion				✓
Accept previous other matter			✓	✓
Authority to proceed to next stage		✓	✓	✓

Progressive architect's information requirements

Table 7.13 Progressive architect's information requirements.

	ARCHITECT'S INFORMATION	Estimate	CP1	CP2	CP3
Building use		✓			
i	**Design**				
	Sketches	✓			
	• GA plans	✓	✓		
	• GA elevations	✓	✓		
	• GA sections	✓	✓		
	• Landscaping			✓	
	• Functions				✓
	• Construction details elevations				✓
	• Construction details sections				✓
	• Interfaces			✓	
	• Locations				✓
	• Assembly				✓
	• Components				✓
	• Room concepts				✓
	• Constraints			✓	
ii	**Floor areas/accommodation**	✓	✓		
iii	**Outline specification**	✓			
	• Elements		✓		
	• Quality		✓		
	• Components, materials			✓	
	• Acoustics				✓
	• Outline performance criteria		✓		
	• Finishes				✓
	• Alternative specifications			✓	

Table 7.13 (*Continued*)

	ARCHITECT'S INFORMATION	**Estimate**	**CP1**	**CP2**	**CP3**
iv	**Room data sheets**				✓
v	**Fittings schedule**				✓
vi	**Strategies**				
	• Environmental	✓			
	◦ BREEAM			✓	
	◦ Building Regulations		✓		
	◦ Sustainability				✓
	◦ Renewable energy			✓	
	◦ Specific requirements				✓
	• Car parking	✓			
	• Vertical movement		✓		
	• IT			✓	
	• Fire			✓	
	• Acoustics				✓
	• Security				✓
	• DDA			✓	
	• Window cleaning		✓		
	• Refuse			✓	
	• Public art				✓
	• Conservation		✓		
	• Sundry				✓
vii	**Reports**				
	Archaeological		✓		
	Topographical			✓	
viii	**Construction methodology**				✓
ix	**Fit out**				✓
x	**Risk**				✓

Progressive services information requirements

Table 7.14 Progressive services information.

	M&E SERVICES INFORMATION	Estimate	CP1	CP2	CP3
i	**Drawings**				
	• GA		✓		
	• Schematic		✓		
	• Plant room			✓	
	• Diagrams				✓
	• Layouts				✓
ii	**Specification**	✓			
	• Mechanical			✓	
	• Electrical			✓	
	• Transport				✓
	• Protective				✓
	• Communications			✓	
	• Security				✓
	• BMS		✓		
	• Special				✓
	• Plant schedule		✓		
	• Duties			✓	
	• BWIC		✓		
	• Alternatives				✓
iii	**Strategies**				
	• Environmental		✓		
	• BREEAM			✓	
	• Building Regulations		✓		
	• Sustainability				✓
	• Renewable energy			✓	

Table 7.14 (*Continued*)

	M&E SERVICES INFORMATION	Estimate	CP1	CP2	CP3
	• Specific requirements				✓
	• Vertical movement		✓		
	• Removals				✓
	• Commissioning			✓	
iv	**Reports**				
	• Existing services		✓		
	• Abnormals			✓	
v	**Utilities**				
	• Connections	✓			
	• Upgrading		✓		
	• Diversions		✓		
	• Quotations				✓
vi	**Facilitating works**				✓
vii	**Risk**				✓

Progressive structural information requirements

Table 7.15 Progressive structural information.

	STRUCTURAL INFORMATION	Estimate	CP1	CP2	CP3
i	**Reports: desktop**				
	• Contamination		✓		
	• Geotechnical		✓		
	• Bombs		✓		
ii	**Reports: fieldwork**				
	• Contamination			✓	
	• Geotechnical			✓	
iii	**Environmental risk**				✓

(*Continued*)

Table 7.15 (*Continued*)

	STRUCTURAL INFORMATION	Estimate	CP1	CP2	CP3
iv	**Ground conditions**	✓			
v	**Design**				
	• GA		✓		
	• Frame		✓		
	• Layout				✓
	• Sections			✓	
	• Foundations		✓		
	• Substructure sections		✓		
	• Drainage		✓		
vi	**Levels**		✓		
vii	**Specification**	✓			
	Elements				
	• Components				✓
	• Materials				✓
	• Loadings				✓
	• Piling		✓		
	• BWIC services			✓	
	• Steelwork			✓	
viii	**Reinforcement**				✓
ix	**Steelwork**		✓		
x	**Methodologies**				
	• Facilitating works		✓		
	• Temporary works				✓
	• Alterations				✓
	• Drainage				✓
xi	**Risk**				✓

8 Preliminaries, Risk, Overheads and Profit

8.1 INTRODUCTION

We can now examine the preliminaries which are a substantial proportion of the cost of any project. Like the quantities for the building works the preliminaries require accurate and consistent measurement and pricing to provide an accurate estimate for the costs. NRM 1 introduces a comprehensive schedule of costs to be included and many items also have a method of measurement. This enables the cost of the preliminaries to be transparent from estimate to cost plan to tender and across different contracts. In due course, the NRM schedule of preliminaries could be adopted by the contractors for tendering purposes. The introduction of several categories of risk also requires a calculation. This chapter suggests methods of fulfilling these requirements, using simple risk impact analysis calculations.

The quantity surveyor usually calculates preliminaries as a simple percentage of the anticipated contract sum. The contractors usually have their own standard schedule of preliminaries which is completed for each contract. Many disputes concerning the valuation of variation and the final account could be resolved more easily if an accurate and consistent set of comprehensive preliminaries is available.

8.2 PRELIMINARIES

The preliminaries are shown in NRM 1, Part 4, Element 9, pp. 277–306. This indicates what is to be measured. This is another good example of standardisation and consistency.

There is a choice concerning the equipment for site preliminaries. Should they be purchased or leased? Generally most of this equipment is leased. This is because capital

Estimating and Cost Planning Using the New Rules of Measurement, First Edition. Sean D.C. Ostrowski.

expenditure requires cash-in-hand that is best used in the production process itself, which can generate profits, rather than the expenditure on administration which is an overhead expenditure. The purchase would occur before the income stream starts and would require cash-in-hand or an overdraft. In addition, the erection and dismantling of plant and equipment generates substantial removal and storage costs. Also, refurbishment costs are considerable and include transport and reassembly. The purchase, finance costs, maintenance and refurbishment cost of plant and equipment make a leasing arrangement cost-effective.

Although the NRM indicates what should be measured there are different methods of measurements for different types of preliminaries. The general rule is that measurement provides more accuracy than a percentage and that quotations from specialists are better than educated guesses. A detailed spreadsheet of each item is required to measure preliminaries. An example is given in Table 8.1.

Table 8.1 Sample spreadsheet layout for measuring preliminaries.

FUNCTION	SIZE	DURATION	RATE	TOTAL
Accommodation				
Site manager	$20\,m^2$	100 wks	£3/m^2	6,000
Meeting room	$50\,m^2$	100 wks	£3/m^2	15,000
Alterations/repairs				
			Allowance	10,000
Cleaning	2 hrs	100 × 5 days	£10/hr	10,000
Consumables		100 wks	£100/wk	10,000

An extract from the comprehensive schedule set out in NRM 1 for management and staff, is shown in Table 8.2.

Table 8.2 Preliminaries example.

Element 9.2: Main contractor's cost items

Sub-element 9.2.1: Management and staff

Component	Included	Unit	Excluded
1 Project specific management and staff	Main contractor's project specific management and staff such as:		1 External design consultants (included in group element 12: Project/design team design fees).
	1 Contractor's project manager.	Man hours per week	
	2 Construction manager.	Man hours per week	2 Security staff (included in sub-element 10.2.4: Security).
	3 etc.		
etc.	etc.	etc.	etc.

Source: RICS (2012, p. 281). Reproduced by permission of the RICS.

There are several areas that require some additional information to improve the extent of measurement and accuracy. Some items are set out below.

Samples

A considerable proportion of work on site is now prefabricated. Samples are therefore necessary to ensure compatibility at the interface between trades and standards of workmanship. The relevant subcontractors should provide a quotation for the provision of samples during the design stage and immediately prior to commencement on site.

Commissioning

Commissioning is carried out by the most expensive labour on site, usually in restricted spaces with pressing deadlines to achieve. Costs need to include premium rates, hotel costs, travel expenditure and bonuses. Quotations from the subcontractors will provide more accurate prices.

Commissioning fuel will be required during the latter part of the contract for 24 hours per day and 7 days per week. Fuel consumption for each building can be provided by the services engineer.

Consumables/expenses

Management contracts can include large numbers of off-site meetings which generate considerable expenditure for travel and subsistence.

Scaffolding

The cost of scaffolding is usually a quotation from a subcontractor. However, the cost of adaptations and additional hire can increase the cost substantially. Sufficient allowances for adaptations and additional hire should be included for each part of the scaffolding quote.

Small plant

Battery-driven tools and hand-held tool are often provided in standard proprietary containers from specialist plant hire companies. Popular tools often run out and replacements are very expensive. Suitable allowances for replacements should be added to the quotation.

Security

Security on site usually becomes ineffective after 18 months due to complacency on the part of the subcontractors' staff, who are often casual agency personnel. Allowances should be included in the costs for regular replacement of security companies.

The use of swipe card systems is quite commonplace. However, storage and retrieval of the data is not usually included in the quotation. Suitable allowances should be included in the preliminaries for adequate record keeping and retrieval.

Cleaning

Allowances in the preliminaries are usually for a 'builders' clean prior to handover. However, the client's representatives usually require a higher level of cleanliness, the 'sparkle' clean standard. Again suitable allowance should be included in the preliminaries for several cycles of cleaning.

Snagging

Work to be remedied during the defects liability period should be carried out by the relevant subcontractor. However, this is often multitask work and requires addition work from a specialist snagging team. This requires further allowances in the preliminaries.

The period just before and just after practical completion can mean the damage of protective covers as they are repeatedly removed and replaced. Further allowances for additional protection are required in the preliminaries.

Attendances

Most subcontracts require additional attendances from the main contractor to clean and maintain work areas and the site. Further allowances for additional attendance are required in the preliminaries.

Lost and stolen

Although insurance is an overhead, lost and stolen items may come below the excess threshold of most all-risk construction insurance policies. This is particularly the case for the final fix of electrical systems and sanitary ware after the trade handover and before the practical completion. Allowances for direct replacement costs should be included in the preliminaries.

Access

Usually, changes to the access requirements can neither be anticipated nor calculated. However, allowances for these changes should be included.

Hoarding

Site hoarding is now extensively used for advertising purposes. The cost of the hoarding can be offset against advertising revenue or can be undertaken directly by the advertising company.

Catering

On-site catering is subsidised by the main contractor to ensure that breaks and meals are procured without excessive loss of time caused by going off site for refreshments.

Traffic management schemes

Traffic management schemes require temporary traffic lights for 24 hours per day, 7 days per week. Fuel consumption for a petrol generator requires a person and a van to service the generator. The cost of this servicing far outweighs the hire cost of the equipment.

Calculations of preliminaries and temporary works

Suggestions on how to calculate these figures are included in Tables 8.3–8.5.

Table 8.3 Calculation of preliminaries and temporary works.

DESCRIPTION	INFORMATION	CALCULATION	
PRELIMINARIES			
STANDARD TERMS			
Tendering alterations	Ajusting errors/adding new information	Estimate of cost/day	
Increased costs	NEDO/BCIS/estimate	Extrapolate indices	
Fees	DS fees	Published tables/provisional sum	
	Health and safety	Estimate/provisional sum	
Samples	Preambles	Subcontractor quote	
	Assembly	Estimate	
Commissiong fuel	Assess time period	Estimate	
Taxes	Landfill tax	Percentage	Provisional sum
	VAT	New work	0%
		Brownfield	5%
		Refurbishment	17.5%
Insurances	Contractors' all risk	Percentage	
	Employers' liabilities	Percentage	
	Joint refurbishment	Percentage	
MANAGEMENT			
Detailed schedule of personnel × time period		Estimate	
Number of meetings		Estimate	
Extent of travel		Estimate	
Hotels/expenses		Estimate	
PLANT			
Detailed schedule or each item x time period		Quote	
Alterations and adjustments		Estimate	
Small site plant		Estimate	
CONSUMABLES			
Detailed schedule or each item × time period		Estimate	
EXPENSES			
Detailed schedule or each item × time period		Estimate	
SECURITY			
Agree specification with architect		Quote	
HANDOVER/COMMISSIONING			
Preambles for procedures. Specialist gangs × time period		Estimate	
Premium rates/additional hours/hotels/travel		Estimate	

Table 8.4 Calculation of preliminaries and temporary works.

DESCRIPTION		INFORMATION	CALCULATION	
SUBCONTRACTORS				
	Remedials/snagging		Included in quote	
	Attendances/clearing up. Number of gangs × time period		Estimate	
WARRANTIES				
	Performance bond	Percentage of contract sum	Quote	Provisional sum
	Collateral warranties	Percentage of contract sum	Quote	Provisional sum
	Subcontractor warranties	Percentage of subcontract sums	Quote	Provisional sum
RISK REGISTER				
	Schedule of at risk items	Assessed percentage risk. Cost of each delay		Provisional sum
OVERHEADS		Particular to the contract	Percentage	
		Regional	Percentage	
		Group/HQ	Percentage	
PROFIT		Published accounts/directors' review		
		Strength of order book	Percentage	

Table 8.5 Calculation of preliminaries and temporary works.

DESCRIPTION	INFORMATION	CALCULATION
TEMPORARY WORKS		
ACCESS		
Roads	Hardcore and removal	Measure for rate/m^2
Parking	Tarmac and removal	Measure for rate/m^2
Unloading	Heavy duty h/c and removal	Measure for rate/m^2
Alterations		Estimate for lump sum
FENCING	Hoarding	Measure for rate/m
SITE HUTTING		
Prepare detailed schedule of accommodation for purchase or rent		Measure m^2 for quote
Heating and power		Service provider quote
Furniture and fittings	Purchase or rent from suppliers schedule or rates	
	Short-term rental or reuse existing	
Telephones/power/water	Can be calculated per month	Quote from suppliers
Catering/subsidies		Quote from suppliers
SCAFFOLDING		Quote from suppliers
Adaptations	Cost per lift × No. of changes	Estimate
Skips		Estimate
SECURITY	Lighting and alarms	Quote from suppliers
SERVICES	Diversions	Service provider quotes
UNDERPINNING/BUTTRESSES	Specification from consultant	Quote from subcontractor
PROTECTION OF WORK	To be included in subcontractors' quote	
After practical completion	Meaure surfaces per m^2	Estimate
TRAFFIC CONTROL	Agree spec'n with architect	Quote from subcontractor
DEWATERING	Assessment of plant/mth	Estimate

8.3 RISK

In the past, contingencies have been included in estimates in order to provide an assessment of risk. The quantity surveyor faces two opposing factors when assessing contingencies. First, the amount of contingency should reflect the risk involved. This is a combination of the extent of the information that is available and the anticipated performances of the full project team, including the client. Secondly, a contingency figure above a modest percentage is usually perceived as a reflection on the accuracy of the estimate. Anything over 2.5% will usually trigger this response. These mutually irreconcilable requirements have not been resolved.

Contingencies have been replaced in the NRM 1 by risk and there are now four different areas of risk with a schedule of the elements for each. This is Group element 14 (NRM 1, p. 249). These specific calculations are intended to provide transparency when assessing the risks involved in capital projects. Their strengths and weaknesses will be examined in turn.

Risk workshops provide the opportunity to provide mutual appraisal of risk between the teams of consultants. The need to provide a more formal appraisal of risk will require additional resources and time for their preparation. The requirements for successful risk management include: risk registers, options, appraisals, ranking, expected financial impact, assessments and simulation. Data analysis can be quantitative (eg Monte Carlo analysis), qualitative or semi-quantitative. The acquisition of the technical competencies in the use of these tools is a prerequisite of successful risk calculations. A simple example of a semi-quantitative matrix is included in the section on construction risks later in this chapter. Progressive risk calculations which develop from the estimate through each of the cost plans have the advantage of demonstrating to the client the adjustment in risk as the design develops.

Design development risk

The first risk is the design development risk. This is the extent to which the design will change. Several possible areas of risk are listed in Table 8.6.

Table 8.6 Design development risk.

Element 13.1: Design development risks
1 Inadequate or unclear project brief
2 Unclear design team responsibilities
3 Unrealistic design programme
4 Ineffective quality control procedures
5 Inadequate site investigation
6 Planning constraints/requirements
7 Soundness of design data
8 Appropriateness of design (Constructability)
9 Degree of novelty (i.e. design novelty)
10 Ineffective design coordination
11 Reliability of area schedules
12 Reliability of estimating data: • Changes in labour, materials, equipment and plant costs; and • Inflation (i.e. differential inflation due to market factors and/or timing)
13 Use of provisional sums (i.e. do not give price certainty)

Source: RICS (2012, p. 322). Reproduced by permission of the RICS.

The risk for design changes is highest at the early stages and the risk is proportionate to the extent that the design team have worked with each other and with the client.

Matters are further complicated by the calculation itself. The assessment is to consider the risk to a particular element of the design eg coordinated design, and to establish the impact that this might have on the total cost of the building works. For example, a simple percentage for the design development risk multiplied by the cost of the building works would generate a significant financial impact as follows:

Design development risk		Building works		Cost of risk
Say 5%	×	£1,000,000	=	£50,000

There are some fundamental problems here. The first is that there is always some risk. This means that even a minimal percentage on each of the element items in Group Element 13 will result in a significant percentage. The second is the impact that this will have on the working relationships with the designers because a poor appraisal of the anticipated performance of any part of the design team is likely to prejudice the working relationship. Finally, the declaration of such risk may prejudice the viability of the project itself. To overcome these problems a joint spreadsheet can be prepared for discussion by all parties, including the client, so that the assessment of risk is a communal and agreed undertaking.

Construction risks

The construction risk is the extent to which the work will change due to problems with the construction process. The fundamental risk is to do with the programme. This can be substantially reduced with a resourced programme. Element 13.2 (Table 8.7) lists substantial possible areas of risk.

Table 8.7 Construction risks.

Element 13.2: Construction risks
1 Inadequate site investigation
2 Archaeological remains
3 Underground obstructions
4 Contaminated land
5 Adjacent structures (ie requiring special precautions)
6 Geotechnical problems (eg mining and subsidence)
7 Ground water
8 Asbestos and other hazardous materials
9 Invasive plant growth
10 Tree preservation orders.
11 Ecological issues (eg presence of endangered species)
12 Environmental impact
13 Physical access to site (ie restrictions and limitations)
14 Existing occupancies/users
15 Restricted working hours/routines
16 Maintaining access
17 Maintaining existing services
18 Additional infrastructure
19 Existing services (ie availability, capacity, conditions and location)
20 Location of existing services
Etc.

Source: RICS (2012, p. 322). Reproduced by permission of the RICS.

Employer change criteria risks

The employer change criteria risk is the extent to which the client will change the employer's requirements. The fundamental risk is the extent to which the client's brief is comprehensive. The experience of the client is usually an indicator of the extent of likely change. This can be substantially reduced with an appropriate procurement strategy that allows the early participation of the whole team. However, it is unlikely that substantial changes can be anticipated if they are unknown to the client. For instance, the need for blast proof glazing to the Ministry of Defence procurement headquarters in Bristol, UK after the commencement of work was due to the threat caused by increased terrorist activities. Element 13.3 (Table 8.8) lists possible areas of risk.

Table 8.8 Employer's risk change criteria.

Element 13.3: Employer change risks
1 Specific changes in requirements (ie in scope of works or project brief during design, pre-construction and construction stages)
2 Changes in quality (ie specification of materials and workmanship)
3 Changes in time
4 Employer-driven changes/variations introduced during the construction stage
5 Effect on construction duration (ie impact on date for completion)
6 Cumulative effect of numerous changes

Source: RICS (2012, p. 323). Reproduced by permission of the RICS.

As with the design risk, this risk can only be properly assessed with an extensive series of calculations which may not be appropriate, or even possible, at the earlier stages of the project.

Employer's other risks criteria

The employer's other risks are a further range of problems that might occur, such as: funding, planning delays or constraints, sectional completion, problems with designers and management fees. Table 8.9 lists possible areas of substantial risk.

Table 8.9 Employer's other risks.

Element 13.4: Employer's other risks

1 **Project brief:**
- End user requirements
- Inadequate or unclear project brief
- Employer's specific requirements (eg functional standards, site or establishment rules and regulations, and standing orders)

2 **Timescales**
- Unrealistic design and construction programme
- Unrealistic tender period(s)
- Insufficient time allowed for tender evaluation
- Contractual claims
- Effects of phased completion requirements (eg requesting partial possession)
- Acceleration of construction works
- Etc.

Source: RICS (2012, p. 324). Reproduced by permission of the RICS.

Risk analysis

The risk can only be properly assessed with an extensive series of calculations which may not be appropriate or even possible at the earlier stages.

Risk analysis is a complex operation which has some sophisticated mathematical models in use. Whether it is appropriate to use them at this stage at this stage is problematical. A simple semi-quantitative probability and impact analysis table may be more useful to assist in assessing risk at this stage. The probability of an event on a risk register needs to be assessed and then the extent on the impact needs to be assessed to provide an aggregate impact. A simple example follows.

Table 8.10 Semi-quantitative risk analysis.

PROBABILITY AND IMPACT TABLE							
			IMPACT				
			Very Low	**Low**	**Medium**	**High**	**Very High**
			<35%	**45%**	**55%**	**65%**	**>75%**
PROBABILITY	**Very Low**	**<35%**					
	Low	**45%**					
	Medium	**55%**					
	High	**65%**					
	Very High	**>75%**					

Source: Derived from Ashworth (2010), reproduced by permission of Allan Ashworth.

To illustrate how to use this we will consider the possibility of the impact of delays to the programme because of changes to the internal layouts. NRM 1 lists this item in Section 13.2 Construction risks, item 36, Cumulative effect of numerous changes/variations on the construction; Section 13.3 Employer change risks, item 1 Specific changes in requirements (ie scope of the works or project brief during the design, pre-construction or construction stages); Section 13.4 Employer other risks, item 1, Project brief: end user requirements.

If the probability of the risk of changes to the internal layouts is considered to be high, say 65%, the assessment of the impact might be considered to be even higher, say greater than 75%. The risk of delay to the contract would be an aggregate of these two. This is shown in Table 8.11.

Table 8.11 Probability and impact tables.

PROBABILITY AND IMPACT TABLE							
			IMPACT				
			Very Low	**Low**	**Medium**	**High**	**Very High**
			<35%	**45%**	**55%**	**65%**	**>75%**
PROBABILITY	**Very Low**	**<35%**					
	Low	**45%**					
	Medium	**55%**					
	High	**65%**					**Delays due to internal layouts**
	Very High	**>75%**					

Source: Derived from Ashworth (2010), reproduced by permission of Allan Ashworth.

The risks of changes to the internal walls and partitions has therefore been assessed as very high ie > 75%, say 80%. But this is only applicable to the room layouts not to the circulation corridors, which is 50% of the total. This set out in Table 8.12.

Table 8.12 Risk calculation.

Cost of partitions	**Room partitions**	**Design risk**		**Cost of risk**
1,200,000		× 80%	=	960,000
960,000 × 50%	= 480,000	× 80%	=	480,000
Cost of risk	**Total cost of building works**			
480,000	÷ 17,500,000		=	**2% Total risk**

Although this will provide transparency for all parties the extent of the calculations at this stage makes the assessment of these risk factors expensive and time consuming.

Information

This brings us to the schedule of information required for estimates (NRM 1, p. 20) and cost plans (NRM 1, p. 53 and Appendix F, p. 352). They have been examined in detail in Chapter 7. If these information requirements are fulfilled the risk would be reduced substantially. To a large extent the risk is proportional to the extent to which the information is provided in advance of the preparation of the estimates and cost plans.

8.4 OVERHEADS AND PROFIT

Overheads can accrue at local, divisional and international levels and all should be included in the costs for the particular site. However, if the costs of overheads are not completely covered in the site specific tender they need to be recovered elsewhere. As with plant marginal costing can defer the cost of all the overheads to the head office. This is particularly the case with site financing charges which may be deferred to other sites or later time periods.

Profit levels are available from published accounts where profit on turnover can be ascertained. However, the amount of profit to be included in a particular tender is dependent on the strength of the order book and the directors' view of the market.

Contracts that are project management, design and build or cost plus have subcontractor packages that include both preliminaries, overheads and profit that accrue to the subcontractor. The management package will also have preliminaries and profit but they are unlikely to reflect the full extent of either. Several of these packages will have preliminary allowances that are for the total contract.

It is unlikely that the analysis of the preliminaries, overheads and profit will be provided by the subcontractor, who may have no contractual relationship with the client.

8.5 PRACTICAL EXAMPLE: SITE BASED PRELIMINARIES

A typical example of site based preliminaries is set out in Table 8.13.

Table 8.13 Site preliminaries.

PRELIMINARIES Site wide period: 30 weeks					**Date:**	**2005**
Description		**Qty**	**Unit**	**Rate**	**£**	**£**
PLANT	Safety inspections	6	Nr	250	1,500	
	Road sweeper	26	Weeks	600	15,600	
	Sundry plant	1	Item	3,000	3,000	20,100

Table 8.13 Site preliminaries (*Continued*)

PRELIMINARIES Site wide period: 30 weeks					Date:	2005
Description		**Qty**	**Unit**	**Rate**	**£**	**£**
ACCOMMODATION	Hoarding (move and erect)	180	m	60	10,800	
	Hutting 300m^2 × £17.5/m^2/p.a.	26	Weeks	101	2,625	
	Erect		Item		10,000	
	Dismantle		Item		5,000	
	Office cleaning	26	Weeks	175	4,550	
	Computer equipment	26	Weeks	75	1,950	
	Photocopier	26	Weeks	50	1,300	
	First Aid equipment and dressings	1	Item	500	500	
	Toilet consumables	26	Weeks	25	650	37,375
SERVICES	Telephones	26	Weeks	200	5,200	
	Temporary electrics	26	Weeks	200	5,200	
	Connection charge for temp. water	1	Item	2,050	2,050	
	Water fountain	26	Weeks	30	780	13,230
STAFF	Producion manager	26	Weeks	1,750	45,500	
	Site manager for shell and core	26	Weeks	1,300	33,800	
	Site manager for fit out	26	Weeks	1,300	33,800	
	Assistant site manager	26	Weeks	1,150	29,900	
	Planner 80%	26	Weeks	1,400	36,400	
	Typist/site clerk	26	Weeks	650	16,900	196,300
SKIPS	Hire of skips	78	Nr	115	8,970	8,970
SCAFFOLDING	Adaptions only				10,000	10,000
PROTECTION	Units	-	Nr	Nil	Nil	Nil
LABOUR	Street cleaner	26	Weeks	475	12,350	
	Site cleaning labourer	26	Weeks	475	12,350	
	Welfare labourer	26	Weeks	475	12,350	
	Welfare labourer	26	Weeks	475	12,350	

Table 8.13 Site preliminaries (*Continued*)

PRELIMINARIES Site wide period: 30 weeks					**Date:**	**2005**
Description		**Qty**	**Unit**	**Rate**	**£**	**£**
	Agency supervisor	15	Visits	525	7,875	
	Handyman (sales centre)	26	Visits	750	19,500	
	Handyman	26	Weeks	750	19,500	96,275
SMALL TOOLS	Small tools	26	Weeks	50	1,300	1,300
CONSUMABLES	Protective clothing	10	Nr	150	1,500	
	Fire precautions	1	No	1,200	1,200	
	Progress photographs	6	No	250	1,500	
	Consumables stationery	26	Weeks	50	1,300	5,500
SECURITY	Security gate (day)	26	Weeks	800	20,800	
	Security gate (night)	26	Weeks	800	20,800	
	Induction	20	Weeks	800	16,000	
	Security central traffic area	20	Weeks	800	16,000	73,600
SIGNAGE	Notices/signs	1	Item	3,500	3,500	3,500
	SITE MANAGEMENT BUDGET TOTAL				**466,150**	**466,150**

8.6 SELF-ASSESSMENT EXERCISE: WEEKLY RUNNING COSTS

Calculate the weekly cost for the 30-week contract period.

Compare your own work with the proposed solution included in Appendix 9 (Appendix 9 is on the website (http://www.wiley.com/go/ostrowski/estimating)).

Self-assess your work on the assessment sheet included in Appendix 4 (Appendix 4 is on the website (http://www.wiley.com/go/ostrowski/estimating)).

To provide further assistance there are dedicated websites at http://ostrowski quantities.com and at Wiley Blackwell (http://www.wiley.com/go/ostrowski/estimating). It is hoped that the provision of this will go some way towards explaining the concepts and principles more clearly than using the printed word alone.

9 Unit Rates

9.1 INTRODUCTION

The measurement of the estimate now requires pricing. The price for any measured work requires the following:

- Cost of labour, rate of pay
- Time necessary to do works, labour constant
- Cost of plant, rental cost
- Time necessary for plant to do work, work rate
- Cost of the material
- Amount of material that is necessary
- Preliminaries necessary to carry out work
- Overheads that management requires
- Profit level to be achieved

Estimating and Cost Planning Using the New Rules of Measurement, First Edition. Sean D.C. Ostrowski.

Much of this is provided by subcontractors submitting tenders for work at the invitation of the main contractor. The prices in these tenders are the rates for each unit of measure, but they do not provide the analysis of labour, materials and plant. The subcontractors' submissions are 'cover' prices for the work. They enable the main contractors to price the work with reference to the prices from the subcontractors. This enables the tenders to include two important matters. First, the tenders will incorporate the prices that are available from a subcontractor. Secondly, the work can be carried out in the proposed programme period by experienced subcontractors.

The ability to build up these prices is a core competence for quantity surveyors. It enables the prices to be accurately estimated. In turn, this provides a check on the prices submitted by the subcontractors. The analysis of these prices also provides the quantity surveyor with the amount of labour, plant and materials necessary to carry out the complete contract. This is the basis of a resourced programme that is necessary for the management team to provide an accurate assessment of the time to be taken to carry out the works. At the post-contract stage the final account includes large numbers of variations which require accurate pricing. The ability to break down the rates given in the bill of quantities to their constituent parts enables an accurate price for the variation to be built up. In addition, subcontractors' submissions on the pricing of these variations can be easily checked if it is possible to build up the proposed rate from the basic components of labour, materials and plant costs.

9.2 LABOUR RATES

RICS prime cost of daywork

The first part that is required is the cost of the labour. The basic hourly rates applicable to all labour are negotiated each year by representatives of unions and employers and are published as the Working Rule Agreement by the Building and Allied Trades Joint Industrial Tribunal. The 2010 craftsman rate was £10.62/hr and the general operative rate was £9.88/hr. These rates of pay have been incorporated into the prime cost of labour applicable to labour in the UK and are published by the RICS as the 'Definition of Prime Cost of Daywork Carried Out Under a Building Contract' 3rd edition, 2007. These are the mandatory rates applicable to work done using standard form of contract.

The contractor may use other rates. The bills of quantities usually include a section for pricing a provisional sum item for dayworks. This is work that cannot be measured and is priced by recording the labour, materials and plant used to carry out the works as follows:

Allowances for provisional sum and percentage addition required for all additional costs including overheads and profits (O/H & P) are added to the prime cost to form a separate section of the tender documents.

Dayworks

LABOUR				
Craftsmen		100 hrs £13.33		1,333.00
General operative		200 hrs £9.88		1,976.00
				3,309.00
Add	Percentage addition required	say 150%		4,964.00
				2,000.00
MATERIALS				
Add	Percentage addition required	say 15%		300.00
				1,000.00
PLANT				
Add	Percentage addition required	say 10%		100.00
	PROVISIONAL SUM		TOTAL	£11,673.00

The percentage additions are for the additional management and administration for carrying out the dayworks and for the price adjustments to the labour for the actual rate paid for the work. There is a wide range of percentages on differing contracts, from 25% to 250%, depending on the circumstances and the type of work.

The RICS Prime Cost of Daywork provides the reference for all rates for measured work and for the rates in 'measured term' contracts and in daywork accounts. Most disputes concerning the cost of labour can be resolved by reference to these calculations.

The intention is that these are the minimum rates for labour and that they are applicable to all labour on site. This provides a standardised pricing structure that is accurate and consistent. The decline of the unions, the introduction of the 'minimum rate', the National Minimum Wage Act (1998) and the National Minimum Wage Regulations (1999), the influx of labour from around the world and the extensive use of agencies for the provision of labour have reduced the extent to which these rates are generally used. The minimum wage rate in 2010 was £5.93. An example of a build up of the Prime Cost of Daywork using the current Working Rule Agreement is given in Table 9.1.

Table 9.1 Hourly rate build up using RICS prime cost of dayworks.

LABOUR RATES										
DESCRIPTION					CRAFTSMAN		LABOURER			
52 × 39 − (163 + 63) = 1,802hrs ÷ 39 = 46.2wks										
Rates at 13.9.10					Rate			Rate		
					£	£		£	£	
Guaranteed minimum weekly earnings										
	Standard Basic Rate	39hrs × 10.62/hr = £414.18	46.2	wks	414.18	19,135.12	39 × 7.88	307.32	14,198.18	
Employer's National Insurance Contribution		(414.18 − 110) × 12.8% × 46.2				1,798.80			1,166.87	
Employer's contribution to:-										
	Holiday pay	226 × 10.62				2,400.12	226 × 7.88		1,780.88	
	Welfare benefit	52 × 11.00				572.00			572.00	
	CITB levy @ 0.50% of payroll	(19,135.12 + 2,400.12) × 0.5%				107.68			79.90	
						24,013.71			17,797.83	
Hourly cost of labour as defined in section 3 Clause 3.02						**£13.33**			**£9.88**	
24,013.71 ÷ 1,802										
How many working days per year ? 52 × 5 = 260 − 22 − 8 = 230dys		Most estimators use 220 working days available per year								

9.3 LABOUR CONSTANTS

Having established the cost of labour, it is then necessary to establish how long it will take to do the work. The time taken to build the work is based on the labour constants for that particular trade. These constants have been established for some considerable time and are used across the whole range of construction activities and by the publishers of pricing books. They vary in accordance with the skill and output of the operative and the weather and site conditions.

They are called constants because they remain reasonably accurate as an average. Two examples will illustrate this. Manual excavation of a pad foundation not exceeding 1.50 m deep takes 5 hours per cubic metre. Laying and jointing cast iron drainage pipes in trenches remains the same, whatever the depth of the trench or the location. These labour constants have been used in England and Hong Kong to verify the prices and labour requirements.

New technology, particularly prefabrication, requires new labour constants. For example the fixing of curtain walling is essentially the attendance of a labour gang to manoeuvre the prefabricated glazing panel into place from the tower crane and around the scaffolding and the fixing of the panel to prepared bolts in the structure. These labour requirements are constant and applicable worldwide.

Labour constants are provided in the well-known publication by W. Atton *Estimating Applied to Building* (Atton, 1975). An extract for brickwork is given in Table 9.2.

Table 9.2 Labour constants.

Common bricks		
Labour constants per m^2		
Description	**Skilled hours**	**Unskilled hours**
102.5 mm thick below damp proof course (DPC)	0.90	0.45
215 mm thick below DPC	1.80	0.90
327.5 mm thick below DPC	2.50	1.25
Reduced brickwork below DPC	1.60	0.80
102.5 mm thick above DPC	1.00	0.50
215 mm thick above DPC	2.00	1.00
327.5 mm thick above DPC	2.70	1.35
Reduced brickwork above DPC	1.80	0.90
102.5 mm in filling old openings	1.10	0.55
215 mm in filling old openings	2.20	1.10
327.5 mm in filling old openings	3.00	1.50
Reduced brickwork in filling old openings	2.00	1.00
102.5 mm in skin of hollow walls	1.00	0.50
215 mm in skin of hollow walls	2.00	1.00

Source: Atton (1975, p. 78).

Pricing books also contain labour constants. The cost of the labour part of the price is the rate per hour multiplied by the labour constant. In most cases the labour cost that is published is the result of this calculation. There is usually a calculation of the rate per hour at the beginning of the book but rarely is the labour constant published. We can examine a typical example of a pricing book rate (Table 9.3) to establish the labour constant.

Table 9.3 Brickwork labour rates.

Spon 2006 1B facings (PC £275/100) in Flemish bond £88.00/m²				
Labour constant 2.08 hr/m²				
A gang of 13 men (1 F/M + 6 B/L + 4 Lab) gives 6.5 m² an output of per hour In this gang there are 6 bricklayers In a normal gang of 3 men (2 B/L and 1 Lab) the output is 2 m² per hour The output of 2 bricklayers providing 2 m² per hour is very similar to 6 bricklayers providing 6.5 m² per hour (2.08 × 3 gangs = 6.24 m² per hour)				
Labour (in hours)	**Labour cost in £**	**Materials**	**Total**	
2.08	45.25	43.15	88.42	
Foreman	**Bricklayer**	**Labourer**		
£13.17/hr	£12.44/hr	£9.31/hr		
Gang size				
1 Foreman	1 × 13.17	13.17		
6 Bricklayers	6 × 12.44	74.64		
4 Labourers	4 × 9.31	37.24		
		125.05	÷ 6.5 m²/hr	= 19.24
			+ 13% OH & P	= 21.76
Labour cost in Spon 2.08 × 21.76 = 45.25				

Source: Davis Langdon (2011).

The rates per hour for foreman (F/M) (£13.17), bricklayer (B/L) (£12.44) and labourer (Lab)(£9.31) are the Spon calculation of the RICS Prime Cost of Labour.

9.4 MATERIALS

Materials are often purchased in bulk and measured in different units. For example, concrete materials of aggregate, sand and cement are purchased by weight but measured by volume. However, for estimates concrete beams and columns are measured by the square metre, for cost plans they are measured in linear metres and for bills of quantities they are measured in cubic metres. For plasterboard partitioning the metal studwork is purchased in linear metres, the plasterboard in square metres and the measurement for the estimates and cost plans is in square metres.

The conversion calculations are also constants that are published. Again most of these constants are provided by Atton (1975). An example for brickwork is given in Table 9.4.

Table 9.4 Materials constants.

Bricks	
The following is the number of bricks required per square metre of wall per 102.5 mm thickness of wall, with 10 mm thick cross and bed joints	
50 × 102.50 × 215 mm bricks	74 number
65 × 102.50 × 215 mm bricks	59 number
75 × 102.50 × 215 mm bricks	52 number
90 × 102.50 × 215 mm bricks	50 number
Example	
The following is the number of 65 × 102.5 × 215 mm bricks with 10 mm cross and bed joints required for a 327.5 mm thick wall, per m^2	
327.5 mm wall = 3 No. 102.5 mm thicknesses + wall joints	
= 3 × 59 = 177 bricks	
Number of bricks required = 177 bricks per m^2	
Waste	
Add 5% on common bricks	
Add 2.5% on engineering bricks	
Add 5% on mortar	

Source: Atton (1975, p. 77).

Waste is a continuous problem. The waste allowances that are published are indicative at best and are rarely sufficient. However, including the full cost of waste in the calculation of the price will cause the individual rate to be excessive, although it will be factual. This is overcome by including the full allowances for the purchase of additional materials that are caused by waste, loss or theft in the overheads.

9.5 PLANT

The price of the plant comprises the cost of the plant and the output that can be achieved from each piece of machinery. Again the output of the various kinds of plant is published and the main outputs are shown in Atton (1975).

Most plant is hired and the following commentary relates mainly to hired plant. The rates for plant hire vary for the following reasons:

- Local vagaries of supply and demand
- Substantial discounts are available but are dependent upon volume of business
- Actual costs are distorted by allowances for depreciation
- The working life depends on use/abuse of each machine
- Extent of maintenance
- Utilization: the appropriate use for the function, eg mini diggers in restricted spaces means that premium rates are applicable where the utilisation is limited
- Extent of utilisation should be based on actual records rather than published information. For instance, the time taken to transport machinery between sites is not included in output capacities
- Obsolescence: the plant is only obsolete if it cannot be used. For instance, very occasional use or old plant does not necessarily mean cheap rates are available for the hire of the plant
- Replacement costs require a sinking fund which includes allowances for inflation and high rates of interest
- The use of full absorption costing or marginal costing by the plant hire company can affect the hire rate
- Running costs, and sometimes insurance, need to be added to the hire costs

The use of plant is an increasing proportion of the cost of the work and is reflected in the prices that are tendered. The changes are summarised annually in the Plant Hire Investment Report by C. Stratton. The two major changes are the increase in the amount of plant being used in relation to the total cost of construction and the change in ownership from construction companies to dedicated hire companies (Table 9.5).

Table 9.5 Changes in plant hire 1993–2003.

	CONSTRUCTION SPEND	GROSS BOOK VALUE OF PLANT	%	PERCENTAGE HELD BY PLANT HIRERS
1993	17,492,000,000	1,300,000	7%	15%
2003	29,643,000,000	3,639,000,000	12%	60%

Source: Stratton (2003). Reproduced by permission of Catherine E. Stratton.

The reasons for the relative increased cost of plant are:

- Increasing availability of useful machinery. This is due to cross-hiring between hire companies
- Decrease in the use of manual labour, which is a less efficient means to move materials
- Impact of the Health and Safety at Work Acts
- Increasing cost of plant as prices rise due to less competition

The reasons for the movement from construction company-based plant to dedicated plant hire companies are:

- Capital cost of plant restricts trading opportunities for contractors
- Maintenance management and costs are transferred into hire costs
- Capital liability cost of hired plant does not appear on balance sheet
- Cross-hiring from plant hire companies provides a much larger pool of available plant at any one time

The advantages of the correct use of plant hire can be considerable. The following example shows a more than ten-fold reduction in the cost of the work if a machine can be used instead of manual labour to dig a trench. However, the work needs to be in sufficient quantity to justify the hire of the machine, the machine has to have access to the site and access to the workface and the hire rate needs a substantial discount. The constants are shown for labour and machine output (Table 9.6).

Table 9.6 Comparison of machine and hand excavation rates.

MANUAL EXCAVATION	
Labour constant for excavating surface trench 1.50–3.00 m deep	3.250 hr/m^3
Labour constant for staging platforms. One per 1.50 m high	1.525
Labour constant for clearing back to top	1.525
Total labour constant per m^3	**6.500**
Rate per m^3 for hand excavation 6.50 hr @ £10/hr	**£65.00/m^3**
MACHINE EXCAVATION	
Mini excavator with 0.25 m^3 bucket in stiff clay	7 m^3/hr
Hire of 0.8 m^3 mini excavator £150/day × 50% discount	75.00
Driver	100.00
Transport to and from site	135.00
Cost per day	310.00
8 working hours per day × 7 m^3/hr = 56 m^3 per day	
Cost per day = £310.00 ÷ 56 m^3 per day = £5.54/m^3	
Rate per m^3 for machine excavation	**£5.54/m^3**

9.6 PRACTICAL APPLICATION: FOR CONCRETE, BRICKWORK, PARTITIONING, ROOFING, WINDOWS

Concrete

Some unit rates are given in Tables 9.7 to 9.10. They are provided with commentary as appropriate.

Table 9.7 Concrete unit rate calculation.

CONCRETE						
Plain in-situ hand mixed concrete mix C20, 21 N/mm², ratio 1:2:4, 20 mm aggregate, poured against earth or unblended hardcore; 150 mm thick for foundations to footpaths or the like						
MATERIALS						
	Cement @ £110.00/t, sharp sand @ £14/t, aggregate @ £16/t					
	Cement	1 m³	=	1,200 kg		
	Sharp sand	1 m³	=	1,800 kg		
	Aggregate	1 m³	=	1,600 kg		
	Cement	$\frac{1,200}{1,000}$	× £110/t	=	132.00	
	Sharp sand	$\frac{1,800}{1,000}$	× £14/t × 2	=	50.40	
	Aggregate	$\frac{1,600}{1,000}$	× £16/t × 4	=	102.40	
	1 part cement				284.80	
	2 parts sand		Add voids @ 40%		113.92	
	4 parts aggregate				398.72	
	7 parts in total		Divided by 7 parts		÷ 7	
					56.96	
			+ waste 5%		**59.81**	per m³
PLANT						
	10/7 mixer @ £50/wk					
	2 m³ per 6 minutes					
	20 m³ per 60 minutes					
	2 men mixing @ £10.00/hr		= £20.00			
	2 men carrying @ £10.00/hr		= £20.00			
		Hire	$\frac{£50.00}{40\ hr}$	=	1.25	
		Labour	Mixing		20.00	
			Carrying		20.00	

Table 9.7 *(Continued)*

CONCRETE continued						
					41.25	per 20 m³
			Divided by 20		÷ 20	
					2.06	per m³
LABOUR						
	Foundations @ 4 hr/m³		4 hr @ £10.00		**40.00**	per m³
TOTALS			Materials		59.81	
			Plant		2.06	
			Labour		40.00	
					101.87	
			O/H & P 10%		10.19	
				£	**111.06**	
Non-productive time for plant is not included because it is not predictable. Allowances are included in the overheads. The percentage for overheads and profits increases as the work becomes more complex.						

Brickwork

Table 9.8 Brickwork unit rate.

BRICKWORK					
One brick thick wall in facing bricks in stretcher bond and gauged mortar (1:1:6)					
MATERIALS	Cement @ £110/t Hydrated lime @ £120/t Building sand @£15/t Facing bricks @ £300/1000	**PLANT**		10/7 mixer £50.00/week	
CONSTANTS	Cement	1 m³	=	1,200 kg	
	Lime	1 m³	=	600 kg	
	Building sand	1 m³	=	1,800 kg	
Voids	25%				
Volume of mortar per m²	1B wall 0.06 m³ per m²				
Number of bricks	1B wall in facings in stretcher bond 120/m²				
10/7 mixer	2 m³ in 6 minutes				

(Continued)

Table 9.8 (*Continued*)

BRICKWORK continued						
Hours per m² for gang of 2 bricklayers and 1 labourer						
Bricklayer	2 hr/m²					
Labourer	1 hr/m²					
MATERIALS	Cement	$\frac{1,200}{1,000}$	× £110/t	=	132.00	
	Lime	$\frac{600}{1,000}$	× £120/t	=	72.00	
	Sand	$\frac{1,800}{1,000}$	× £15/t × 6	=	162.00	
Cement	1 part				366.00	
Lime	1 part		Add voids @ 25%		91.50	
Sand	6 parts				457.50	
Total	8 parts		Divided by 8 parts	÷ 8	67.19	per m³
MORTAR						
Volume of mortar per m²	1B wall 0.06 m³ per m²					
	£67.19 × 0.06 m³ mortar per m³				**4.03**	per m²
BRICKS	1B wall in facings in stretcher bond 120/m²					
			£300 ×	$\frac{120}{1000}$	**36.00**	per m²
PLANT	10/7 mixer £50.00/week					
	2 m³ in 6 minutes					
	20 m³ in 60 minutes					
	2 men mixing × £10.00/hr = £20.00					
	2 men carrying × £10.00/hr = £20.00					
Cost of hire per hour	Mixer	£50	×	$\frac{1\text{ hr}}{40\text{ hr}}$ =	1.25	
	Labour			Mixing	20.00	
				Carrying	20.00	
					41.25	per 20 m³
			Divided by 20	÷ 20	2.06	per m³
				× 0.06 m³ per m²	**0.12**	per m²

c/fwd

Table 9.8 (*Continued*)

BRICKWORK continued								
LABOUR	2 and 1 Gang	Stretcher bond bricklayer 2 hr/m^2 Labourer 1 hr/m^2						b/fwd
			2 No	B/L	@	13.50	27.00	
		1 hr/m^2	1 No	Lab	@	10.00	10.00	
							37.00	per m^2
TOTALS		**MATERIALS**	Mortar			4.03		
			Bricks			36.00		
						40.03		
			Waste @ 5%			2.00	42.03	
		PLANT					0.12	
		LABOUR					37.00	
							79.15	
			O/H & P 15%				11.87	
						£	**91.02**	per m^2

Hardwood windows

Table 9.9 Hardwood window unit rate.

HARDWOOD WINDOW		
Hardwood window with two casements, plugged screwed and pelleted		
Hardwood window 1200 × 600	£179.00	each
Hardwood window cill 25 × 145	£8.29	per lm
75 mm brass hinges	3.55	per pair
Brass casement stay and pin	6.99	each
Brass casement fastener	5.29	each
Brass screws	2.05	per 25
Mastic	2.79	per tube/11 m
Mastic gun	3.10	each
Mortar	1.99	per 5 kg bag

(*Continued*)

Table 9.9 *(Continued)*

HARDWOOD WINDOW continued			
MATERIALS			
Window		179.00	
Cill	$\frac{1,200}{1,000}$ × £8.29/m	9.95	
Hinges	2 × 1½ prs per casement = 6 No. ÷ 2 = 3 pairs × 3.55/pair	10.65	
Stays	2 no. × 6.99	13.32	
Fasteners	2 no. × 5.29	10.58	
Screws	Hinges 3 pairs × 16 = 48		
	Stays 2 × 4 = 8		
	Fasteners 2 × 4 = 8		
	Total 64 × $\frac{64}{25}$ × 2.05	5.25	
	Waste 5% × 5.25	.26	
Bedding Mortar	Say 0.05 m^3 × $\frac{\text{(Weight) } 1,600}{\text{(Volume) } 1,000}$ × $\frac{1.99}{5}$ = 0.03	Say 0.50	
Mastic	2/1200 2400 2/600 1200 2/3600 7200 Say 8 m; 8 m × $\frac{2.79}{11}$	2.03	
Mastic gun	3.10 ÷ 10 uses	.31	**229.82**
LABOUR: Labour constants from Atton (1975)			
Window	0.80 hr for 1200 × 600 window + 25% for hardwood		
	0.80 hr + 25% = 1 hr × £13.50	13.50	
Casement	0.75 hr per pair of hinges + 25% for hardwood		
	0.75 hr + 25% = 0.94 × 3 pairs × £13.50	38.07	
Cill	0.10 hr/m skilled; 1.20 m × 0.10 × £13.50	1.35	
	0.50 hr/m unskilled; 1.20 m × 0.50 × £13.50	8.10	
Mastic	0.1 hr/m unskilled; 2/1.20 = 2.40		
	2/0.60 = 1.20 3.60		
	3.60 m, say 4 m × 2 = 8 m × 0.18 hr/m × £10.00	13.40	
Stays and fasteners	4 × say 0.50 hr each × 13.50	26.00	**100.42**
TOTALS	Materials	229.82	
	Labour	100.42	
	Allowance for pelleting	5.00	
		335.24	
	O/H & P 15%	50.29	
	£	**385.53**	

Table 9.10 Partitions unit rate.

PARTITIONS

The price per linear metre for proprietary insulated metal stud partition panels of 'Gyproc' or equal and approved metal stud proprietary partitioning comprising 50 mm wide metal stud frame at 600 centres, 50 mm channel plugged and screwed to concrete through 38 × 48 mm 'Tanalised' softwood sole plate: 3.30–3.60 m high.

Tapered edge panels, joints filled with joint filler and tape to receive direct decorations (measured separately) 100 mm partition, 2 hr fire resistant panels, 2 nr. layers of 12.5 mm 'Fireline' board both sides

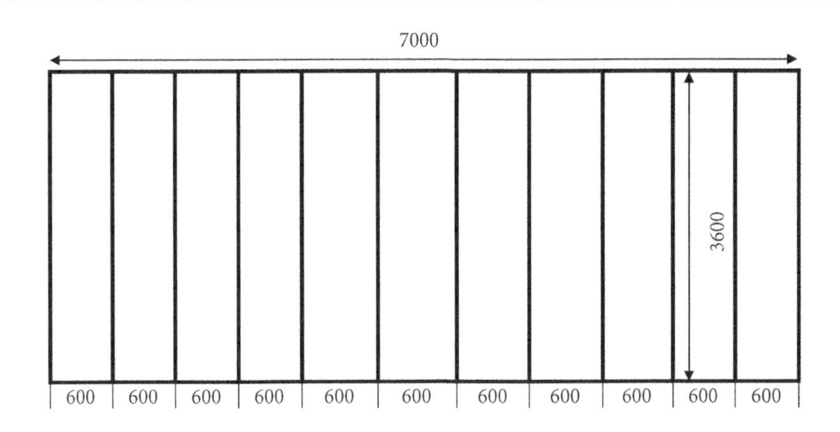

Nominal size 7000 × 3600 = 25.20 m². Say 25 m²
Actual size 11/600 = 6600 + 12/50 = 600
6600 + 600 = 7200 long × 3600 high = 25.92 m²
Prices are calculated for a wall 7000 long × 3600 high and an area of 25 m²
Price is per linear metre as SMM7 or per m² as NRM 2

MATERIAL COSTS
Framework U channel Ref. 72C50 in 3600 lengths @ £8.71 each × 50% discount
Vertical I struts Ref. 70I50 in 3600 lengths @ £4.62 each × 50% discount
Insulation Isowool 50. Packs of 15.60m² @ £6.56 each × 50% discount
12.5 mm Fireline board 1200 × 2400 @ £7.55 each × 50 % discount
Tape 150 m roll @ £3.74 each
Primer/sealer 10 litre tub @ £38.65 each. Coverage 100 m²/15 litres
Sawn softwood 38 × 48 £0.70/m
Screws 25 mm £5.63/1000. Requirement 600 per 25 m²
41 mm £6.93/1000. Requirement 600 per 25 m²

CONSTANTS

Hours per square metre for gang of 1 skilled fitter and 1 labourer for

Dry metal partition 80 mm skeleton	0.30 hr/m²		
12.5 mm board	0.15 hr/m²		
Insulation (1 man only)	0.25 hr/m²		
Primer/sealer	0.30 hr/m²		

(*Continued*)

Table 9.10 (*Continued*)

PARTITIONS continued							
MATERIALS							
Framework channels for perimeter 72C50 in 3600 lengths							
2/3.60	7.20						
2/7.00	14.00	21.20	÷ 3600	= 6 lengths	6 × £8.71 × 50%		26.13
Vertical metal studwork 70I50							
10/3.60	36.00	÷ 3600	= 10 lengths	10 × 4.62 × 50%			23.10
Sole plate				7 m × 0.70			4.90
Insulation Isowool APK 600							
7.00 3.60	25.20 m²	$\frac{25.20}{15.60}$	×	£36.56	× 50%		29.53
12.5 mm Fireline boarding							
4/7.00 3.60	100.80 m²	2400 × 1200 = 2.88 m²/board 100.80 ÷ 2.88 = 35 boards		35 × £7.55 × 50%			132.13
Screws	25 mm	1,000 ÷ 600 = 1.666 × £5.63					9.38
	41 mm	1,000 ÷ 600 = 1.666 × £6.93					11.55
Tape							
	2/2/7.00	28.00					
7000 ÷ 1200	6/2/3.60	43.20	71.20	71.20 ÷ 150 = 0.47 × £3.74			1.78
Sealer							
2/7.00 3.60	50.40 m²	$\frac{50.40}{100}$	= 0.504 × 15 litres	= 7.56 litres	×	$\frac{38.65}{10}$	29.22
							269.69
						Waste 5% including joint filler and edge sealant	13.48
							283.17
						÷ 7 =	**£40.45/m**

(*Continued*)

Table 9.10 (*Continued*)

PARTITIONS continued				
LABOUR				
Partition skeleton	25.20 m^2	× 0.30 hr/m^2	× £13.50 hr	102.06
			× £10.00 hr	75.60
Boards	25.20 m^2	× 0.15 hr/m^2	× £13.50/hr × 4 boards	204.12
			× £10.00/hr × 4 boards	151.20
Insulation	25.20 m^2	× 0.25 hr/m^2	× £10.00 hr	63.00
Sealer	25.20 m^2	× 0.11 hr/m^2	× £10.00 hr × 2 sides	55.44
				651.42
			÷ 7 =	**£93.06/m**

TOTALS					
Labour	651.42 ÷ 25 m^2 =		26.06		93.06
Materials	283.17 ÷ 25 m^2 =		11.33		40.45
			37.39		133.51
O/H & P 15%			5.61		20.03
		per m^2	**43.00**	**per m**	**£153.54**

Roofs

The build up of the price for roof tiling includes the construction and terminology which is common with this type of roof. Diagram 9.1 indicates standard top hung tiles and a normal double overlap.

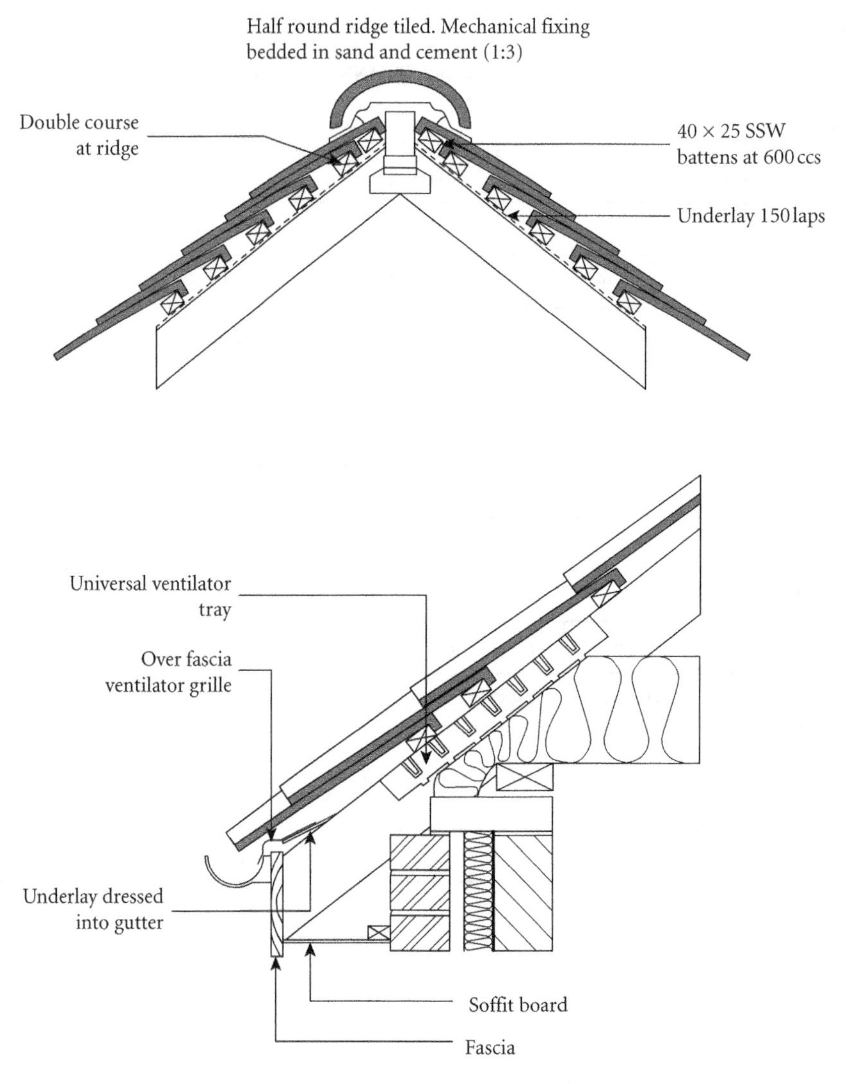

Diagram 9.1 Roof tiling.

The calculation for the number of tiles per square metre (Table 9.10) depends on:

- Size of tile
- Single or double overlap
- Interlocking tiles, pantiles, have single overlap but also side overlap
- Size of overlap
- Top hung or centre hung. The calculation below is for a 265 × 165 top hung tile with 65 mm overlap at the top and no side overlap. The length of the tile minus the lap is divided by two to deal with the double overlap

Table 9.11 Calculation for number of roof tiles per square metre.

$1\,m^2 \div \left[\dfrac{(\text{Length of tile} - \text{lap})}{2} \times \text{width of tile}\right]$
$1.00 \div \left[\dfrac{(265-65)}{2} \times 165\right]$
1.00 ÷ 100 × 165
1.00 ÷ 0.100 × 0.165
1.00 ÷ 0.0615
= 60 tiles

Table 9.12 Roofing tiles unit rate.

ROOFING				
Plain clay tiles, natural sand faced, £750/1000, size 265 × 165 mm, 65 mm laps, nailed every 5th course with 12 gauge × 38 mm aluminium composite nails, 25 × 40 mm pressure impregnated battens fixed to rafters at 400 mm centres, reinforced PVC underlay 1000 gauge with150 end laps and 75 mm side laps.				
MATERIALS Plain clay tiles, 265 × 165 mm, natural sand face £750/1000 Treated ssw 25 × 38 × 2400 long. £8.19/pack of 8 lengths Reinforced underlay1000 mm × 10 roll £28.99 Nails 40 mm £2.29 per bag of 500g weight bag at 300 nails per kg				
LABOUR CONSTANTS				
Hours per square metre for gang of 1 roofer and 1 labourer				
	Plain tiles	Battens	Sarking felt	
Roofer	1.50 hr/100 tiles	2.50 hr/100 m		
Labourer	0.75 hr/100 tiles	1.25 hr/100 m	$0.06\,hr/m^2$ + 15% > 50° slope	
MATERIALS				
Tiles	$1\,m^2 \div$ [Length of tile – (lap)] × width			
	$1.00 \div \dfrac{(265-65)}{2} \times 165$			
	1.00 ÷ (0.100 × 0.165)			£ p
1.00 ÷ 0.0615 = 60 tiles + 2.5% waste = 62 tiles		62 ×	$\dfrac{750}{1,000}$	46.50

(*Continued*)

Table 9.12 (*Continued*)

ROOFING continued						
Nails for tiles						
62 tiles / 5 courses	×	2 = 25	25 / 300	×	(£2.29 × 2) = £0.38 × 1.05 waste	0.40
Battens						
62 tiles × 165 mm wide	= 10.23 lm/m^2	×	1.05 waste	= 10.74 m		
£8.19 ÷ (8 × 2400)	= £0.43/m	×	10.74			4.62
Nails for battens 10.74 ÷ 400 c/s = 29 × 2 per rafter = 58		58 / 300	×	(£2.29 × 2) = £0.38 × 1.05 waste		0.93
Underlay		75 mm side laps		150 mm end laps		
		1000 − 75 = 925		10000 − 150 = 9850		
0.93 9.85	= 9.16 m^2/roll	£28.99 / 9.16	=	3.16/m^2	+ 5% waste	3.32
Nails for underlay					Say	0.23
					Materials	56.00
LABOUR						
Tiles	1.50 hr/100 tiles		1.5 hr × 62/100		× £13.50/hr	12.55
	0.75 hr/100 tiles		0.75 hr × 62/100		× £10.00/hr	4.65
Battens	2.50 hr/100 m		2.5 hr × 10.74/100		× £13.50/hr	3.62
	1.25 hr/100 m		1.25 hr × 10.74/100		× £10.00/hr	1.34
Underlay	0.06 hr/m^2	0.06 × £10 hr	£0.72	+ 15% for > 50° slope		0.69
					Labour	22.85
TOTALS					Materials	56.00
					Labour	22.85
						78.85
					O/H & P 15%	11.83
					£	**90.68**

9.7 SELF-ASSESSMENT EXERCISE: REINFORCEMENT

Calculate the cost per tonne for reinforcement as described in Table 9.13.

Compare your own work with the proposed solution included in Appendix 10 (Appendix 10 is on the website (http://www.wiley.com/go/ostrowski/estimating)).

Self-assess your work on the assessment sheet included in Appendix 4 (Appendix 4 is on the website (http://www.wiley.com/go/ostrowski/estimating)).

To provide further assistance there are dedicated websites at http://ostrowski quantities.com/ and at Wiley Blackwell (http://www.wiley.com/go/ostrowski/estimating). It is hoped that the provision of this will go some way towards explaining the concepts and principles more clearly than using the printed word alone.

Table 9.13 Reinforcement rate build up.

<table>
<tr><th colspan="4">REINFORCEMENT</th></tr>
<tr><td colspan="4">12 mm diameter bar to BS4449 hot rolled deformed high yield steel bars Grade 460 including links and spacers and cutting, bending and fixing on site with tie wire at each intersection in foundations</td></tr>
<tr><td colspan="4">MATERIALS
12 mm diameter BS 4449 Grade 460 high yield steel bars £400.00/t
Ties £20/t
Spacer allowance £25/t
Links allowance 5%</td></tr>
<tr><td colspan="4">CONSTANTS</td></tr>
<tr><td colspan="3">Hours per tonne for 12 mm bar in foundations</td><td>Materials</td></tr>
<tr><td></td><td>Cutting and bending</td><td>Fixing</td><td rowspan="3">Tie wire 12 kg/t</td></tr>
<tr><td>Fixer</td><td>25 hr/t</td><td>30 hr/t</td></tr>
<tr><td>Labourer</td><td colspan="2">Unloading and stacking 2 hr/t</td></tr>
</table>

10 Cost Analyses

10.1 INTRODUCTION

The BCIS use a specialist and technical vocabulary. Some of this vocabulary is reproduced below.

Category: A specialist kind of work eg steelwork, concrete.
Cost: The cost paid for the labour, plant and material.
Cost planning: Anticipated costs of the contract before commencement.
Element: A particular coherent part of the building eg substructure or the roof. They are made up of several different components and trades. For the substructure this includes groundworks, concrete, reinforcement, formwork, piling etc.
Estimate: Anticipated cost of the contract at inception.
Index: Measures changes in the costs or prices of the work.
Price: The complete price to carry out all the work comprising labour, materials, plant, overheads and profit.
Tender analysis: Historical cost analysis after contract completion.
Trade: A particular and separate kind of craftwork eg brickwork, carpentry.

The intended outcome of all the effort to measure and price work correctly is an accurate price for the anticipated construction work. The next step is to collate all the costs for different types of buildings to be able to anticipate the price of future work. Many attempts have been made to create such a crystal ball to enable prices to be forecast.

Estimating and Cost Planning Using the New Rules of Measurement, First Edition. Sean D.C. Ostrowski.

The requirements for this are:

- Collection of the actual cost of previous work carried out
- A system to collate this information into a standard format
- A rolling programme to update the data
- A series of calculations that enable this information to be modified and applied to each new anticipated contract
- A review of the strengths and weaknesses of such a system

A systematic approach to this process is needed to overcome the established perceptions of the industry concerning cost centres. For example it is generally considered that air conditioning is more expensive than gas-fired central heating to radiators. However, air conditioning to open plan offices is cheaper than central heating to partitioned offices. This is because the smaller office spaces require more radiators and pipework, larger plant and more complex building controls. Another example is the relative importance of materials in estimates, which is often overlooked. A normal contract with standard proportions of labour, materials, overheads and profit requires only a relatively modest saving in materials to provide an increase in the profit margin.

10.2 TYPES OF INDICES

Indices are the anticipated prices for the elements of work in an estimate. They are obtained by taking the bills of quantities that is in trade order, as set out in the SMM, from a contract and rearranging the prices for the work into elemental order, as set out in the NRM. When several bills of quantities (BQs) have been treated in this manner it is possible to provide an average price for any particular element. There are three main types of indices in construction.

Input Cost Index: It measures the change in the cost of labour, materials and plant. The data for the basic labour costs comes from nationally agreed labour rates between employer and union organisations which are published annually by the Building and Allied Trades Joint Industrial Tribunal. The actual cost of the labour will reflect the nature of the work, the calibre and workload of the subcontractor and the scarcity of labour. It may not reflect the national agreements. The basic materials and plant cost come from the published rates of a range of supplier organisations. The rates do not reflect the discounts available to substantial customers. The BCIS publish several of these indices in the *BCIS Quarterly Review of Building Prices.* The BCIS Building Cost Index (BCI) uses the nationally agreed labour rates and the published rates from suppliers. They are applied to a standard building model developed by the BCIS which includes weighting to the costs to provide an average building. The BCIS has a general index and several subsets of specialist categories.

Tender Price Index (TPI): This measures changes in the tender prices. These tendered prices are abstracted from the bills of quantities that are supplied to the BCIS and applied to a standard schedule of rates developed by the BCIS for an average building. The standard schedule of rates does not include services which make up 25–50% of most buildings. The standard schedule of rates is for Joint Contracts Tribunal (JCT) standard lump sum contracts. Much work now uses the design and build contract. In 2010 this was nearly 40% of the value of contract work (RICS/Davis Langdon, 2010). The TPI is for a limited number of work categories and contract types.

Output Cost Index: This measures the anticipated changes to the TPI in the future in relation to anticipated changes in the volume of construction work in the future. This is a derivative of the previous indices and is useful for the long-term analysis of large infrastructure and Price Fluctuation Index (PFI) work where cash flow is significant.

The fluctuations in inflation are measured with another kind of cost input index. The 'NEDO' indices measures changes construction prices of labour, materials and in all the major categories. They are now published by the BCIS as Price Adjustment Formulae Indices (PAFI). These indices are used to monitor the inflation rate in construction and where contracts have a clause allowing the contract sum to be adjusted in accordance with the NEDO/PAFI index.

There are two significant matters that should be remembered. The first is the process of breaking down the trade prices and building them up as elemental prices. This is laborious, time consuming and subject to personal interpretation concerning the demarcation between the assignment of items. For instance, drainage below the slab is included in the substructure, but outside the building drainage is included in the external works. The requirements of such a system are similar to those of any standard method. They are standardisation, accuracy and consistency.

The second matter is that it is important to remember that these are not costs. They are the prices submitted by the contractor in the tender. What the costs are will emerge during the contract as the work is carried out. The regular cost reconciliation exercises carried out by the contractor will establish the difference between the income provided in the prices shown in the BQs and the actual expenditure for labour materials and plant. This information is commercially sensitive and is rarely available outside the contracting organisation. This also means that the prices in the BQs do not translate into accurate elemental costs.

These problems are fundamental and bring into question whether or not these procedures are appropriate for the provision of anticipated rates for work. Despite these difficulties there are several different proprietary cost indices available. The BCIS is part of the RICS and publishes a range of indices and pricing books. Large private practices also publish pricing books. Some of them include labour constants as part of the build-up of the prices. The labour constants that are used are different in each publication. Schedules of rates published by contracting organisations for use in measured term contracts are neither prices nor costs. They are charge-out rates for work which include mark ups for overheads and profit.

The availability of large amounts of computer storage space would suggest that a common store of information would be beneficial to all. However, the costs are commercially sensitive and the information that has been laboriously gathered is jealously guarded and is expertise that is highly prized and protected by intellectual copyright. Even if all the price and cost information was published on a website, so that you would be able to use the information to price your own project, the mere publication of the information does not guarantee its accuracy nor that the information is appropriate to any particular project. By way of analogy, in the legal profession many law cases are published and yet it is not possible to say for certain what the outcome of litigation will be.

10.3 REQUIREMENTS OF INDICES

'Different classes of work call for different methods of cost planning and analysis . . .'
Seeley (1976, p. 179)

Information

To take an example: the parties involved in constructing light industrial units will know very accurate costs of cladding and roofing systems. This means that prices can be provided very quickly for different combinations of industrial units on an industrial estate with the work in the substructure being the only major variable in the specification.

The system developed by the BCIS illustrates the requirements on any index. Gathering the information requires the cooperation of the industry in providing the tender information. The BCIS requires 300–350 tenders spread across the year to maintain the statistical significance of the indices that are published. The information is prices for measured work and requires conversion to elemental prices. The information is the tender information. There may be significant changes to the prices between the tender figure and the contract figure.

Prices in price books are provided by contractors and subcontractors. These prices may be subject to significant discount during tender negation and may not reflect the price at which the work was carried out. Many prices for specialist work may be supplied by only one contractor and may not therefore reflect a market rate, for instance scaffolding and lifts. Some prices are published without the analysis of labour and materials, for instance cladding. The statistical analysis that follows is predicated on accurate information and data input procedures and adjustments need to be made (definitely not to be included), based on the perceived accuracy of the information provided.

Standardisation

The coding of the information is extensive and provides a framework for accuracy and consistency. However, the requirements of a fixed system means that adjustments for particular circumstances may be considerable and may lead to distortion. The need to change the rates in the BQs, which are measured in trade categories, into elemental categories, is the most significant example. The examples in Sections 10.6 and 10.7 illustrate this. The TPI also requires the same series of BQs to be regularly repriced with current rates in order to show the change in the tender prices.

Variables

Having established a data gathering operation and a standard set of indices it is then possible to use them to predict the prices for any new contract in the future. Each new contract will be different and will require adjustments to a number of variables. They are:

- The description of the site
- The site conditions
- The type of contract
- The particular specification
- The market conditions

The BCIS have published indices for different locations, for tender prices, for building costs, for market conditions, for the size of a contract and for the type of contract. The adjustments that are necessary to these indices and the particular changes for the size of the site, the site conditions, the type of contract and the particular specification, require the expertise and experience of the estimator. Using these indices alone will not provide an accurate answer.

Expertise and experience

The price needs to be modified by variables which include: size, the market, functionality, labour, materials and plant costs, fluctuations, speed, type of contract, preliminaries and the time of year. Seeley (1976, p. 178) quite rightly says '*Skilled judgement is therefore necessary when extracting and using cost information . . .*'.

He goes further to suggest that '*Some are opposed to the centralised processing of cost information . . . all the necessary background information can only be secured through close familiarity with the project . . .' (p. 179)*.

The differences between internal and external estimates have already been described in Chapter 1, Section 1.1. The complexity and difficulty involved makes estimating the most difficult job in construction.

10.4 PROBLEMS WITH INDICES

Labour costs

Labour costs use labour constants and standard rates of pay. Labour constants are reasonably well established for most trades. However, new methods of working will mean different constants. For instance, metal stud partitions and plasterboard are different to timber and plaster walls. Prefabricated plumbing units and curtain walling do not have published labour constants although they are well known in subcontractor organisations. Productivity is also variable. The use of agency labour often reduces productivity where supervision on site is often undermined by the agency's requirements. The perceived differences in charges made for labour by the agencies and the actual payment received by the labour also undermines productivity. The standard rates of pay established by the Working Rule Agreement have been undermined by the legislation that set a minimum hourly rate. The payments inevitably fall to the lowest level that is legal.

The international labour market also alters the standard. Fit out contractors specialise in the completion of a building after the shell and core has been completed. They are utilising international labour markets which has seen the emergence of large groups of multi-task workers brought in from outside the EU at considerably reduced pay levels.

For example, plasterers are flown in from Mexico in chartered planes. They are provided with accommodation and work seven days a week on a ten-week contract and are then returned home for a two-week holiday. This rolling programme provides a large amount of constantly available labour at much reduced rates. It also generates its own problems. For instance, there may be language problems for labour from east of the EU because their native tongues are in the Cyrillic alphabet which bears little relationship to European languages. Thus, it may be that neither the instructions nor the drawings are fully understood. Labour rates are therefore very specific to the particular circumstances.

Material costs

The published cost of material is often heavily discounted by large subcontractors who can make large-scale purchases over extended periods of time. Cross holdings of equity between suppliers and subcontractors also provides further discounts in lieu of dividend payments. For instance, large plastering subcontractors often make the cost of materials virtually negligible in their pricing because they are able to obtain such substantial discounts. The cyclical nature of painting subcontractor work, which depends heavily on summer-time school painting programmes also presents problems for the accurate pricing of this work. A large holding of standard paint colours often enables painting subcontractors to ignore the cost of the paint when pricing the work if the order book is low. The materials can be purchased later when their cash flow has improved.

Plant costs

Plant costs usually mean the cost of hiring plant. However, the purchase of plant for a particular site can reduce costs considerably. The replacement costs are then transferred to overheads, thus the true cost of the plant is not reflected in the rates. Discounts available from plant hire companies can be considerable if the market is poor or plant is idle. However, cross-hiring of plant between plant hire companies often ensures that bargains are not passed on to the contractor. The productivity of plant is significant if the plant does not match the requirement specification of the particular site. Infrequent, but necessary, use of plant makes the rates that have to be charged much higher than normal.

Abnormal costs

Each building is unique and construction may require abnormal costs. For instance, substructure work may include piling because of the ground conditions, access and parking requirements may be complex, additional structural work may be required because of adjacent buildings. The experience and expertise of the quantity surveyor will enable the abnormal costs to be excluded from the cost analysis and the model of the anticipated costs.

Risk register

The indices do not reflect risk. The NRM has replaced contingencies with a detailed calculation for several types of risk. Although the risk might be considerable for design and build contracts the risk is not reflected in the rates, rather it is included in the overheads. The problems with these calculations are discussed in Chapter 5 of this textbook.

Overheads

The cost of overheads can be the subject of marginal costing if the equipment is hired and does not appear on the balance sheet. If marginal costs are used, the costs of the overheads revert to the holding company, and thus the site costs are considerably reduced. This allows the prices to be considerably reduced.

Profit

Profit is dependent on market conditions. It can only be provided by the contracting organisations. In difficult market conditions, or to secure the work, it will be bought by using a negative profit figure which actually reduces the tender figure to below the anticipated costs in the hope of some form of recovery in the final account. All figures for profit have a large risk element involved.

10.5 USING INDICES TO ADJUST ESTIMATES

Indices have been prepared that provide some key performance indicators which the BCIS call benchmarks. They enable several adjustments to be made for different price adjustment factors. The following examples use BCIS indices although some of the indices are also published by other sources. They reflect costs and prices for simple general building works using a simple standard contract for a standard contract size of £1,000,000. They do not reflect the prices or costs of small-scale domestic works, complex designs, large-scale buildings and complex contractual arrangements. The illustrations are drawn from the practical application exercise at the end of this chapter. They are:

- **SIZE**
- **LOCATION**
- **INFLATION**
- **MARKET CONDITIONS**
- **TYPES OF CONTRACT**

Size

There are three factors to adjust as the size of the building alters:

1. GIFA
2. Elements
3. Monetary size of the contract

1. Gross internal floor area

The first is changing the floor area of the building will affect the anticipated cost. Changing the floor area from 3,500 m^2 to 3,000 m^2 on a contract sum of ££2,500,000 will have the following effect.

$$\frac{(\text{Size of old building}) - (\text{Size of new building})}{(\text{Size of old building})} \times \text{Contract Sum} = \text{Adjustment}$$

$$\text{GIFA:} \frac{(3{,}500 - 3{,}000)}{3{,}000} \times £2{,}500{,}000 = £416{,}667$$

The modification to the GIFA for each building will modify the anticipated cost. The changes to the areas of the cost-significant elements will also modify the anticipated cost. For instance, increase in the size of the building and an increase in the area of the upper floors and the external walls will also increase the anticipated costs. However, these increases may duplicate each other and careful adjustment is necessary to eliminate any duplication. These adjustments may be problematical if the information is not available and will require the first hand experience of the estimator.

2. Elements

The second is individual elements of each building can vary considerably. For instance the external walls:

$$\frac{(\text{m}^2 \text{ ext. walls in new}) - (\text{m}^2 \text{ ext. walls in old})}{(\text{m}^2 \text{ ext. walls in old})} \times \% \text{ Contract} \times \text{Contract Sum} = \text{Adjustment}$$

$$\text{External walls} \frac{(2{,}680 - 2{,}290)}{2{,}290} \times 9\% \times £2{,}823{,}945 = £43{,}300 \text{ addition}$$

The addition of changes to the floor area and changes to the sizes of the elements would duplicate additional costs. Some adjustment to the calculation using the revised floor area may be necessary where one or more elements are considerably altered. For instance, the roof area may be considerably reduced if the same floor area is contained in a multistorey building rather than a single-storey building.

3. Monetary size of the contract

The BCIS publish an index based on the size of the contract as a key performance indicator known as the Pricing Adjustment Factors. A summary for 2010 is given below.

Average	< £1m	1.07
	£1–2m	1.00
	£2–4m	.96
	£4–10m	.95
	£10–15m	.90
	>£15m	.88

What is the adjustment for a similar building where the original cost is £1.5m and the anticipated cost of the new building is £2.5m?

The anticipated price is higher than £2m so the factor is 0.96. The factor is smaller than unity so the adjustment is a decrease in value:

$$\text{contract sum} \times \text{size factor} = \text{adjustment}$$

$$\text{contract size: } 2{,}500{,}00 \times \frac{(0.96 - 1.00)}{1.00} = (\pounds 100{,}000)$$

LOCATION

The anticipated contract sum is adjusted by the locations index. Regional variations to tenders are published by the BCIS as regional and county factors. The regional diversities in the index are reducing as companies centralise the estimating functions and suppliers use national networks for distribution. A selected summary for 2010 is given in Table 10.1.

Table 10.1 BCIS location factors.

LOCATION FACTORS		
REGIONAL	**COUNTY**	**FACTOR**
YORKSHIRE AND HUMBERSIDE		0.92
	Humberside	0.91
	North Yorkshire	0.95
	South Yorkshire	0.93
WALES		0.96
	Clwyd	0.93
	Dyfed	0.99
	Gwent	0.97
LONDON	Postal districts	1.25

The Bridgend location index is 0.97. The Ripon location index is 0.91. The index is lower so the anticipated price is reduced.

$$\frac{(\text{index for Ripon}) - (\text{index for Bridgend})}{(\text{index for Bridgend})} \times \text{Contract Sum} = \text{Adjustment}$$

$$\text{Location: } \frac{(0.91 - 0.97)}{0.97} \times \pounds 2{,}823{,}945 = (\pounds 174{,}677)$$

The location index shows a considerable premium for contract in London. Currently the index for London postal regions is 1.25. However, by using a London contract as a basis for an anticipated contract in the rest of the UK the index would suggest that there is a considerable reduction in the price. This anticipated reduction should be reviewed with some caution. This is examined in the self-assessment exercise in Section 10.7.

INFLATION

The anticipated contract sum is adjusted by the price fluctuations index. Fluctuations to tenders are published by the BCIS as the TPI. A selected summary is given in Table 10.2.

Table 10.2 BCIS inflation factors: Tender Price Index.

TENDER PRICE INDEX		
1985		100
2000	Average	161
2002	Average	187
2005	Average	224
2006	Average	230
2010	Average	216

One can ask: What is the inflation on the same job that was priced at £2,500,000 in 2002 Quarter 4 (Q4) if it was to be carried out in 2006 Quarter 3 (Q3)?

The 2002 Q4 index is 190. The 2006 Q3 index is 236.

$$\frac{(\text{index for 2006 Q3}) - (\text{index for 2002 Q4})}{(\text{index for 2002 Q4})} \times \text{contract sum} = \text{adjustment}$$

$$\text{Inflation: } \frac{(236 - 190)}{190} \times £2{,}500{,}000 = £605{,}236. \text{ This is } 24\%$$

General building costs

The BCIS publish a general building cost index (BCI) based on the cost of materials. A selected summary is given in Table 10.3.

Table 10.3 BCIS Building Cost Index.

BUILDING COST INDEX		
1985		100
2000	Average	190
2002	Average	204
2005	Average	241
2006	Average	255
2007	Average	267
2008	Average	282
2009	Average	285
2010	Average	296

One can ask: What is the increase on the same job that was priced at £2,500,000 in 2002 Q4 if it was to be carried out in 2006 Q3 using the BCI?

The 2002 Q4 index is 204. The 2006 Q3 index is 255.

$$\frac{(\text{index for 2006 Q3}) - (\text{index for 2002 Q4})}{(\text{index for 2002 Q4})} \times \text{contract sum} = \text{adjustment}$$

$$\text{Inflation: } \frac{(255 - 204)}{204} \times £2{,}500{,}000 = £625{,}000. \text{ This is } 25\%$$

The index for tender prices has risen by 24% in the period 2002–6. The index for building costs has risen by 25% in the period 2002–6.

These indices show almost identical changes. However, the construction industry remains very cyclical and demand changes with the strength of the economy. When the market is strong, tender prices surge ahead of building costs. When the market is poor, tender prices remain static despite increases in costs. Comparing these two indices may provide an indicator for anticipated costs in different market conditions. The BCIS have compared the TPI with the BCI and produced a new index known as the market conditions index.

MARKET CONDITIONS

The index for market conditions is calculated by the BCIS as the TPI divided by the BCI, as shown in Table 10.4.

Table 10.4 BCIS Market Condition Index.

MARKET CONDITION INDEX				
Market Condition Index	=	$\frac{\text{Tender Price Index}}{\text{Building Cost Index}}$		
2000		$\frac{161}{190}$	=	0.85
2002		$\frac{187}{204}$	=	0.92
2005		$\frac{224}{241}$	=	0.93
2006		$\frac{230}{255}$	=	0.90
2010		$\frac{216}{296}$	=	0.73

The index is lower so the anticipated price is reduced.

$$\frac{(\text{index for 2002}) - (\text{index for 2006})}{(\text{index for 2006})} \times \text{contract sum} = \text{adjustment}$$

$$\text{Market conditions: } \frac{(0.90 - 0.91)}{0.91} \times £2{,}823{,}945 = (£31{,}032)$$

The TPI for tender prices is reducing and at the same time the BCI for building costs is increasing. The index for 2010 is 0.73 which indicates a substantial reduction of the tender price in relation to the building costs. Using this information for anticipating would indicate a very difficult market where contractors are absorbing the costs of increased materials and still reducing prices.

However, it is not recommended that this index be used to anticipate the impact of market conditions on future prices. It provides derivative information, not direct data. The formation of a third index derived from the first two only provides an indication of the impact of the market.

TYPES OF CONTRACT

The different types of contract are also compared using different types of tendering. This is published by the BCIS as 'Selection of Contractor'. A summary is given in Table 10.5.

Table 10.5 BCIS Selection of Contractor Index.

SELECTION OF CONTRACTOR INDEX	
Mean = 100	Index
Open competition bills of quantities	99
Selected	99
Two-stage	106
Negotiated	113
Serial	104
Continuation	103
Design and build	
Competitive	105
Negotiated	111
Private finance initiative/PPP	110

The competitive tendered BQs has an index of 99. The competitive design and build contract has an index of 114, thus:

$$\frac{(\text{D \& B index})-(\text{BQ index})}{(\text{BQ index})}\times \text{Contract Sum} = \text{Adjustment}$$

$$\text{Contract type: } \frac{(105-99)}{99}\times £2{,}823{,}945 = £171{,}148$$

The collective use of indices

These individual calculations may be used singly, in part, or in full, as the circumstances and information is available. However, they cannot all be used together to cumulatively build up the cost of the new anticipated contract. This is because there is no interconnectedness or relationship between the indices. If all the results from the use of each of the indices were added together the result would be very inaccurate. A detailed example is included in Section 10.6.

10.6 PRACTICAL APPLICATION: COST ADJUSTMENT FOR CUSTOMER SERVICE CENTRE

A client requires to construct a new consumer service centre, with a gross internal floor area of 3,500 m^2, to be located in Ripon, with a probable tender date of July 2006. It will be procured using a design and build contract using a two-stage selective tendering process with the final contract being negotiated with the selected tenderer.

A similar project was tendered in November 2002 in Bridgend.

Prepare a cost plan for the new project in Ripon by using the information provided below by comparing it with the previous contract in Bridgend.

Ripon contract: substructure GIFA 1,750 m^2
Ripon contract: upper floors GIFA 1,750 m^2
Ripon contract: external walls area 2,290 m^2
Ripon contract: windows and external doors area 2,100 m^2

Table 10.6 BCIS tender analysis of customer service centre.

BCIS		
ELEMENTAL ANAYSIS	**Ref No. 21402**	**Type Ref. A-2-3,080**
Building function	320 offices	
Type of work	New build	
Gross floor area	3,080m^2	
Job title	Customer service centre	
Location	Bridgend	
District	Ogwr	
OS grid reference	SS9080	
Dates	Receipt of tender 27.11.2002	
	Base date 17.11.2002	
	Acceptance 10.12.2002	
	Possession 13.1.2003	
Project details	Two-storey office block together with external walls including block paving, lanscaping, services, drainage and site lighting	
Site conditions	Level car park site with good ground conditions. Excavation above water table. Unrestricted working space and access	
Market conditions	Competitive. Project tender price index 156 on BCIS 1985 Index	
Client	Welsh Development Agency	
Tender documentation	Bills of qantities	
Selection	Competative tender	
No. of tenders	Issued 6. Received 6	
Contract	JCT Private 1998 with CDP	
Contract period (months)	Stipulated 9, offered 9, accepted 9	
Fluctuations	Firm price tender	

Table 10.6 (*Continued*)

BCIS continued					
ELEMENTAL ANAYSIS	**Ref No. 21402**			**Type Ref. A-2-3,080**	
Tender list	£2,823,945				
	£2,827,031	+ 0.1%			
	£2,957,480	+ 4.7%			
	£2,984,594	+ 5.7%			
	£3,236,182	+ 14.6%			
	£3,419,981	+ 21.1%			
Contract breakdown	Measured work		£1,514,389		
	Provisional sums		£215,500		
	Prime cost sums		£665,185		
	Preliminaries		£348,496		
	Contingencies		£80,375		
	Contract sum		£2,823,945		
Accommodation and design features: V shaped two-storey customer design centre with open plan offices, mass concrete fill, RC pad foundations and ground slab, PCC upper floor and stairs, steel frame, felt-covered flat and slate-covered pitched roof, rendered block walls, aluminium curtain walls ansd windows, brise soleil, block partitions, flush doors, plaster to walls, carpet tiles and access flooring, mineral fibre suspended ceiling, sanitary ware, Gas HW central heating, comfort cooling, electric ventilation, electric light and power, lift, lighting protection, fire/intruder alarms, CCTV, BMS					
Basement 0 m^2	Useable area 2,073 m^2		Area of external walls 2,930 m^2		
Ground floor 1,448 m^2	Circulation area 601 m^2		Wall to floor ratio 95.13%		
Upper floor 1,632 m^2	Ancillary area 341 m^2		Ground floor storey height	4.20 m	
Gross floor area 3,080 m^2	Internal partitions 65 m^2		Upper floor storey height	3.600 m	
	Gross floor area 3,080 m^2		Volume	17,806 m^3	

Table 10.7 BCIS elemental analysis of customer service centre.

ANALYSIS										
ELEMENT		TOTAL COST		COST PER m^2		ELEMENT QUANTITY		ELEMENT RATE	PERCENTAGE	
1	**SUBSTRUCTURE**	89,877	89,877	29.18	29.18	1,448	m^2	62.07	3%	3%
2	**SUPERSTRUCTURE**									
2A	Frame	116,626		37.87		3,080	m^2	37.87	4%	
2B	Upper floors	75,065		24.32		1,632	m^2	46.00	3%	
2C	Roof	190,096		61.72		1,948	m^2	97.59	7%	
2D	Stairs	17,288		5.61		4	No.	4,322	1%	
2E	External walls	252,087		81.85		2,290	m^2	110.08	9%	
2F	Windows and external doors	245,712		79.78		640	m^2	383.93	9%	
2G	Internal walls and partitions	56,743		18.42		2,228	m^2	25.47	2%	
2H	Internal doors	62,290	1,015,907	20.22	329.79	80	No.	778.63	2%	36%
3	**FINISHES**									
3A	Walls	45,955		14.92		3,057	m^2	15.03	2%	
3B	Floors	154,053		50.02		2,775	m^2	55.51	5%	
3C	Ceilings	43,749	243,757	14.20	79.14	2,715	m^2	16.11	2%	9%
4	**FITTINGS**	122,381	122,381	39.73	39.73				4%	4%
5	**SERVICES**									
5A	Sanitary	45,512		14.78					2%	

ANALYSIS										
ELEMENT		TOTAL COST		COST PER m^2		ELEMENT QUANTITY		ELEMENT RATE	PERCENTAGE	
5B										
5C										
5D	Water	15,133		4.91					1%	
5E										
5F	Space heating	168,230		54.62					6%	
5G	Ventilation	101,027		32.80					4%	
5H	Electrical	154,250		50.08					5%	
5I	Gas	336		0.11					0%	
5J	Lift	21,625		7.02					1%	
5K	Protection	25,825		8.38					1%	
5L	Communications	6,797		2.21					0%	
5M	Special	10,000		3.25					0%	
5N	BWIC	10,823		3.51					0%	
5O	Attendance		559,558		181.67				0%	20%
6	**EXTERNAL WORKS**									
6A	Site works	212,509		69.00					8%	
6B	Drainage	51,824		16.83					2%	
6C	Services	99,261	363,594	32.23	118.06				4%	13%
7	**PRELIMINARIES**	348,496	348,496	113.15	113.15				12%	12%
8	**RISK**	80,375	80,375	26.10	26.10				3%	3%
	TOTALS	**2,823,945**	**2,823,945**	**916.82**	**916.82**				**100%**	**100%**

Table 10.8 BCIS Price Fluctuations Index.

BCIS PRICE FLUCTUATIONS INDICES					
BASE DATE 1985 = 100					
YEAR	QUARTER	INDEX	SAMPLE SIZE	PERCENTAGE INCREASE ON BASE YEAR	PERCENTAGE INCREASE ON PREVIOUS YEAR
2000	1	158	66	58%	
	2	158			
	3	162			
	4	167			
2001	1	170	82	70%	8%
	2	171			
	3	177			
	4	177			
2002	1	182	56	82%	7%
	2	189			
	3	188			
	4	190			
2003	1	196	64	96%	8%
	2	198			
	3	198			
	4	195			
2004	1	200	72	100%	2%
	2	215			
	3	213			
	4	225			
2005	1	221	45	121%	11%
	2	228			
	3	221			
	4	226			

Table 10.8 *(Continued)*

BCIS PRICE FLUCTUATIONS INDICES					
BASE DATE 1985 = 100					
YEAR	QUARTER	INDEX	SAMPLE SIZE	PERCENTAGE INCREASE ON BASE YEAR	PERCENTAGE INCREASE ON PREVIOUS YEAR
2006	1	228	76	128%	3%
	2	231			
	3	228			
	4	232			
2007	1	239	73	139%	5%
	2	241			
	3	248			
	4	251			
2008	1	250	67	150%	5%
	2	247			
	3	246			
	4	241			
2009	1	222	64	122%	-11%
	2	216			
	3	219			
	4	211			
2010	1	210		110%	-5%
	2	210			
	3	210			
	4	210			

Table 10.9 BCIS Location Factor Index.

BCIS LOCATION FACTORS							
REGIONAL		COUNTY		DISTRICT		FACTOR	
516	YORKSHIRE AND HUMBERSIDE					0.92	
		101	Humberside			0.95	
		87	North Yorkshire			0.99	
		124	South Yorkshire			0.96	
301	WALES					0.96	
		49	Clwyd			0.90	
		34	Dyfed			0.96	
		50	Gwent			0.95	
		22	Gwynedd			0.89	
		51	Mid Glamorgan			0.92	
		22	Powys			0.90	
		43	South Glamorgan			0.90	
		30	West Glamorgan			0.90	

Ripon is in the BCIS region of North Yorkshire. Bridgend is in the BCIS region of South Glamorgan.

Table 10.10 Cost plan adjustments for Ripon.

RIPON COST PLAN ADJUSTMENTS												
ADJUSTMENTS TO GIFA							**Contract sum**		**2,823,945**		**Adjustment**	**Contract sum + adjustment**
GIFA	$\frac{(\text{Ripon} - \text{Bridgend})}{\text{Bridgend}}$	×	Contract sum	=	Adjustment							
	$\frac{(3500-3080)}{3080}$	×	2,823,945	=	385,083						385,083	3,209,028
ADJUSTMENTS TO SIZE OF ELEMENTS												
Element	$\frac{(\text{Ripon} - \text{Bridgend})}{\text{Bridgend}}$	×	Percentage of the contract	=	Add/deduct	=	Percentage	×	Contract Sum	=	Adjustments	
Substructure	$\frac{(1750-1448)}{1448}$	×	3%	=	0.21 × 3%	=	0.61%	×	2,823,945	=	17,226	
Upper floors	$\frac{(1750-1632)}{1632}$	×	2%	=	0.07 × 2%	=	0.14%	×	2,823,945	=	3,954	
External walls	$\frac{(2680-2930)}{2930}$	×	8%	=	0.09 × 8%	=	(0.72%)	×	2,823,945	=	(20,332)	
Windows and external doors	$\frac{(2100-640)}{640}$	×	8%	=	2.28 × 8%	=	18.24%	×	2,823,945	=	515,088	

(Continued)

Table 10.10 *(Continued)*

Ripon cost plan adjustments continued											
ADJUSTMENT FOR SIZE OF CONTRACT										515,935	3,339,880
	$\frac{(\text{Ripon} - \text{Bridgend})}{\text{Bridgend}}$	×	Contract sum						=	Adjustment	
	$\frac{(.96 - 1.00)}{1.00}$	×	2,823,945						=	(112,958)	2,710,987
ADJUSTMENTS FOR LOCATION											
	$\frac{(\text{Bridgend} - \text{Ripon})}{\text{Rippon}}$	×	Contract sum								
	$\frac{(.90 - .99)}{.99}$	×	2,823,945						=	(256,722)	2,567,223
ADJUSTMENTS FOR FLUCTUATIONS											
TDI at time of Bridgend tender Q4 2002 is 190						$\frac{(228 - 190)}{190}$	×	2,823,945	=	564,789	3,388,734
TDI at time of Ripon tender Q3 2006 is 228											
ADJUSTMENTS FOR MARKET CONDITIONS											
Market is .92 on BCIS 2002 Index						$\frac{(.92 - .90)}{.90}$	×	2,823,945		(62,754)	2,761,191
Market is .90 on BCIS 2006 Index											
ADJUSTMENTS FOR DESIGN & BUILD											
Index for competitive tenders and BQs is 99						$\frac{(111 - 99)}{99}$	×	2,823,945	=	342,296	3,166,241
Index for Design & Build and selective tendering is 111											

RIPON COST PLAN ADJUSTMENTS
The size of the building has increased from 3,080 m^2 to 3,500 m^2. An increase of 14%. A 14% increase in the size of the building would increase the anticipated cost to £3,219,297.
Adjusting all the elements has increased the cost by £515,935. An increase of 18%. It may be appropriate to make an additional allowance to the floor area adjustment.
The adjustment using the size of the contract index is less than 2% and is marginal.
The adjustment using the location index is a reduction of 9%.
The adjustment using the inflation index is 20%.
The adjustment using the market condition index is less than 1% and is marginal.
The adjustment using the type of tendering index, from competitive tendering to negotiated design and build, is an increase in price of 12%. This is problematic.
If all these indices were used together the result would be an anticipated contract sum of £4,234,671. This is an increase of 50%.
This demonstrates that all these indices cannot be used together.
The use of these indices either individually or collectively does not guarantee an accurate anticipated price.
What is required is the selective use of one or more of the indices based on the particular circumstances of the proposed contract and the expertise and experience of the quantity surveyor.

10.7 SELF-ASSESSMENT EXERCISE: COST ADJUSTMENT FOR EDUCATIONAL BUILDING

Complete the cost plan adjustments of the educaional building for a similar scheme in York.

Compare your own work with the proposed solution included in Appendix 11 (Appendix 11 is on the website (http://www.wiley.com/go/ostrowski/estimating)).

Self-assess your work on the assessment sheet included in Appendix 4 (Appendix 4 is on the website (http://www.wiley.com/go/ostrowski/estimating)).

An educational authority requires to construct a new educational building, with a GIFA of 4,400 m^2, to be located in York, with a probable tender date of December 2010. It is probable that it will be procured using BQs and competitive tendering. A similar project was tendered in May 2005 in London.

Prepare a cost plan for the new project in York by using the information provided concerning the previous contract in Bridgend that is provided below as follows:

York contract

Substructure GIFA	750 m^2
Ground floor GIFA	300 m^2
Upper floors GIFA	3,350 m^2
Total GIFA	4,400 m^2

See also Tables 10.12–10.15.

York is in the BCIS region of North Yorkshire. The London building is in the BCIS region of London Postal Regions.

To provide further assistance there are also are dedicated websites at http://ostrowskiquantities.com and at Wiley Blackwell (http://www.wiley.com/go/ostrowski/estimating). It is hoped that the provision of this will go some way towards explaining the concepts and principles more clearly than using the printed word alone.

Table 10.11 BCIS tender analysis of educational building.

BCIS LSBU		
ELEMENTAL ANAYSIS	**Ref No.**	**Type Ref.**
Building function	Educational	
Type of work	New build	
Gross floor area	8,225 m^2	
Job title	Educational building	
Location	London	
District	Southwark	
Dates	Receipt of tender 17.5.2005	
	Base date 30.5.2005	
	Acceptance 30.5.2005	
	Possession 1.1.2006	
Project details	Nine storey educational building completely filling an urban site complete with drainage and enabling works	
Site conditions	Good ground conditions. Excavation above water table. Unrestricted working space and access	
Market conditions	Two-stage tender. Project tender price index on BCIS 1985 Index	
Client	LSBU	
Tender documentation	Drawings and specifications	
Selection	Two-stage tender	
No. of tenders	Issued 8. Received 8	
Contract	JCT SBC 2005 with CDP No Quantities	
Contract period (months)	Stipulated 18, offered 15, accepted 15	
Fluctuations	Firm price tender	

(*Continued*)

Table 10.11 (*Continued*)

BCIS LSBU continued				
ELEMENTAL ANAYSIS	**Ref No.**		**Type Ref.**	
Tender list	£17,090,000			
	£16,900,000	– 0.01%		
	£17,500,000	+2%		
	£17,900,000	+2%		
	£18,500,000	+ 8%		
	£19,500,000	+ 14%		
Contract breakdown	Measured work		£12,590,000	
	Provisional sums		£500,000	
	Prime cost sums		£1,500,000	
	Preliminaries		£2,000,000	
	Contingencies		£500,000	
	Contract sum		£17,090,000	

Accommodation and design features: Rectangular nine-storey educational building with lecture rooms, theatres, offices gymnasium and cafeteria, piled foundations and RC slab, frame and upper floors steel and RC stairs, asphalt roof, curtain walling,atrium, metal stud partitions, flush doors, carpet tiles and access flooring, mineral fibre suspended ceiling, heat recovery central heating, ducted ventilation, electric light and power, lift, communications systems, lighting protection, fire/intruder alarms, CCTV, BMS

Basement	$0\,m^2$	Useable area	$6{,}189\,m^2$		Area of external walls $4{,}406\,m^2$
Ground floor	$1{,}260\,m^2$	Circulation area	$1{,}541\,m^2$		Wall to floor ratio 54%
Upper floor	$6965\,m^2$	Ancillary area	$0\,m^2$		
Gross floor area	$8{,}225\,m^2$	Internal partitions	$491\,m^2$		
		Gross floor area	$8{,}225\,m^2$		

Table 10.12 BCIS elemental analysis of educational building.

ANALYSIS										
ELEMENT		TOTAL COST		COST PER m^2		ELEMENT QUANTITY		ELEMENT RATE	PERCENTAGE	
1	**Substructure**	1,200,000	1,200,000	146	146	1,260	m^2	952	7%	7%
2	**Superstructure**									
2A	Frame	1,247,000		152		8,225	m^2	152	7%	
2B	Upper floors	1,372,000		167		6,965	m^2	197	8%	
2C	Roof	176,000		21		1,916	m^2	92	1%	
2D	Stairs	513,000		62		3	No.	171,000	3%	
2E	External walls	2,810,000		342		4,406	m^2	638	16%	
2F	Windows and external doors	243,000		30		504	m^2	482	1%	
2G	Internal walls and partitions	452,000		55		861	m^2	525	3%	
2H	Internal doors	269,000	7,082,000	33	862	242	No.	1,112	2%	41%
3	**Finishes**									
3A	Walls	256,000		38		13,200	m^2	19	1%	
3B	Floors	890,000		108		7,890	m^2	113	5%	
3C	Ceilings	137,000	1,283,000	17	163	2,525	m^2	54	1%	8%
4	**Fittings**	196,000	196,000	24	24				1%	1%
5	**Services**									
5A	Sanitary	76,000		9					0%	
5B	Disposal	170,000		21					1%	

(*Continued*)

Table 10.12 (*Continued*)

ANALYSIS CONTINUED											
	ELEMENT		TOTAL COST		COST PER m^2		ELEMENT QUANTITY		ELEMENT RATE	PERCENTAGE	
5C											
5D	Water		343,000		42					2%	
5E	Heat source		402,000		49					2%	
5F	Space heating		581,000		71					3%	
5G	Ventilation		704,000		86.00					4%	
5H	Electrical		1,218,000		148					7%	
5I	Gas		5,000							0%	
5J	Lift		307,000		37					2%	
5K	Protection		251,000		31					1%	
5L	Communications		122,000		15					1%	
5M	Special		20,000		2					0%	
5N	BWIC		210,000		26					1%	
5O	Attendance		420,000	4,829,000	51	588				2%	28%
6	**External works**										
6A	Site works		Included								
6B	Drainage		Included								
6C	Services		Included								
7	**Preliminaries**		2,000,000	2,000,000	243	243				12%	12%
8	**Risk**		500,000	500,000	61	61				3%	3%
		Totals	**17,090,000**	**17,090,000**	**2,087**	**2,087**				**100%**	**100%**

Table 10.13 BCIS price fluctuations index.

BCIS PRICE FLUCTUATIONS INDICES					
BASE DATE 1985 = 100					
YEAR	**QUARTER**	**INDEX**	**SAMPLE SIZE**	**PERCENTAGE INCREASE ON BASE YEAR**	**PERCENTAGE INCREASE ON PREVIOUS YEAR**
2000	1	158	66	58%	
	2	158			
	3	162			
	4	167			
2001	1	170	82	70%	8%
	2	171			
	3	177			
	4	177			
2002	1	182	56	82%	7%
	2	189			
	3	188			
	4	190			
2003	1	196	64	96%	8%
	2	198			
	3	198			
	4	195			
2004	1	200	72	100%	2%
	2	215			
	3	213			
	4	225			
2005	1	221	45	121%	11%
	2	228			
	3	221			
	4	226			

(Continued)

Table 10.13 (*Continued*)

BCIS PRICE FLUCTUATIONS INDICES					
BASE DATE 1985 = 100					
YEAR	QUARTER	INDEX	SAMPLE SIZE	PERCENTAGE INCREASE ON BASE YEAR	PERCENTAGE INCREASE ON PREVIOUS YEAR
2006	1	228	76	128%	3%
	2	231			
	3	228			
	4	232			
2007	1	239	73	139%	5%
	2	241			
	3	248			
	4	251			
2008	1	250	67	150%	5%
	2	247			
	3	246			
	4	241			
2009	1	222	64	122%	–11%
	2	216			
	3	219			
	4	211			
2010	1	210		110%	–5%
	2	210			
	3	210			
	4	210			

Table 10.14 BCIS location factor index for self-assessment exercise.

BCIS LOCATION FACTORS						
REGIONAL		**COUNTY**		**DISTRICT**		**FACTOR**
516	YORKSHIRE AND HUMBERSIDE					0.92
		101	Humberside			0.95
		87	North Yorkshire			0.99
		124	South Yorkshire			0.96
	LONDON					1.22

Appendix 1

LONDON ROAD DRAWING: NO. SDCO/1/01 SITE LAYOUT, SIZE A1

Estimating and Cost Planning Using the New Rules of Measurement, First Edition. Sean D.C. Ostrowski.

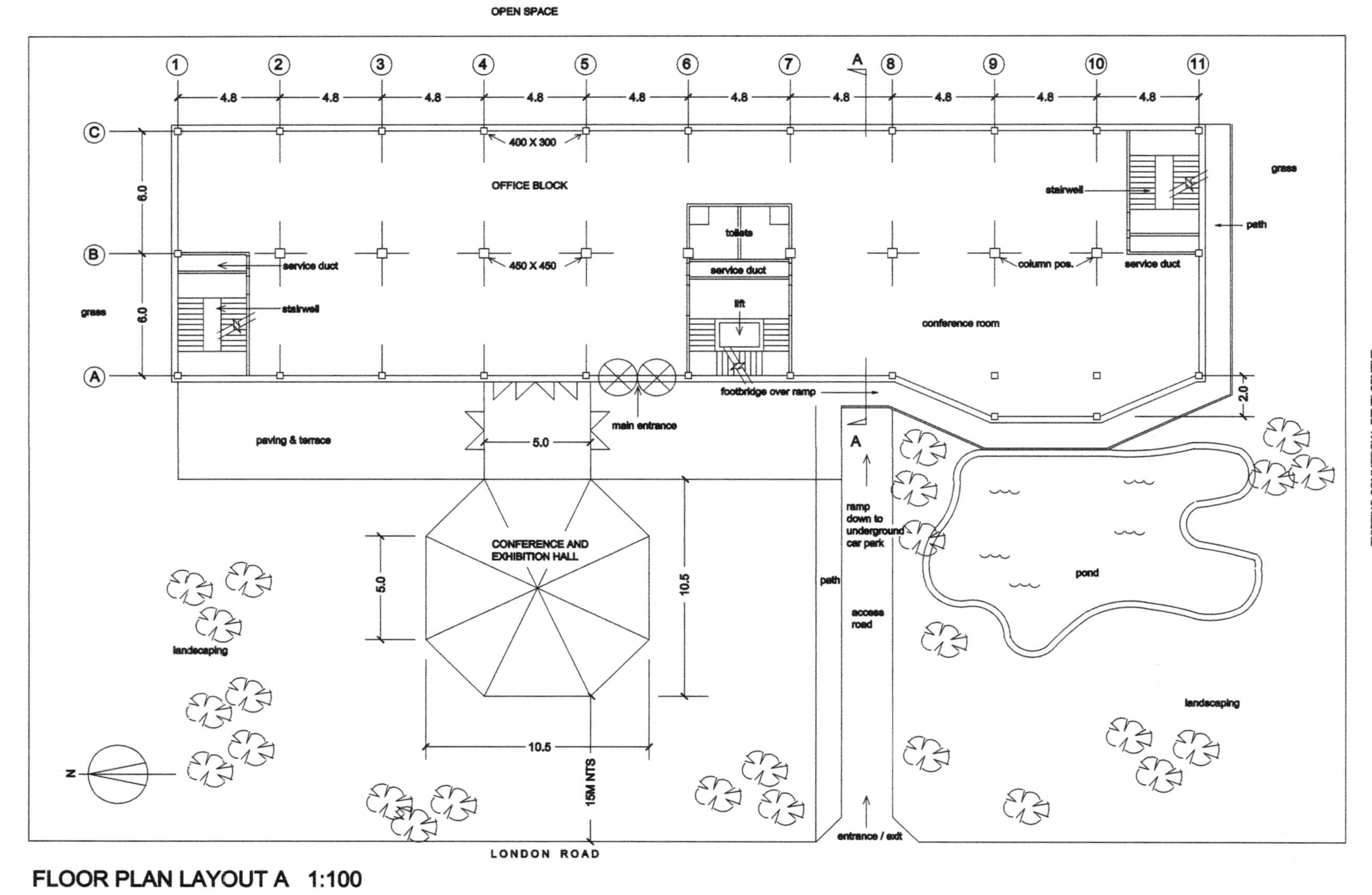

FLOOR PLAN LAYOUT A 1:100

Reproduced by permission of The College of Estate Management.

Appendix 2

LONDON ROAD DRAWING: NO. SDCO/1/02 PLAN, ELEVATION AND SECTION. SIZE A1

Estimating and Cost Planning Using the New Rules of Measurement, First Edition. Sean D.C. Ostrowski.

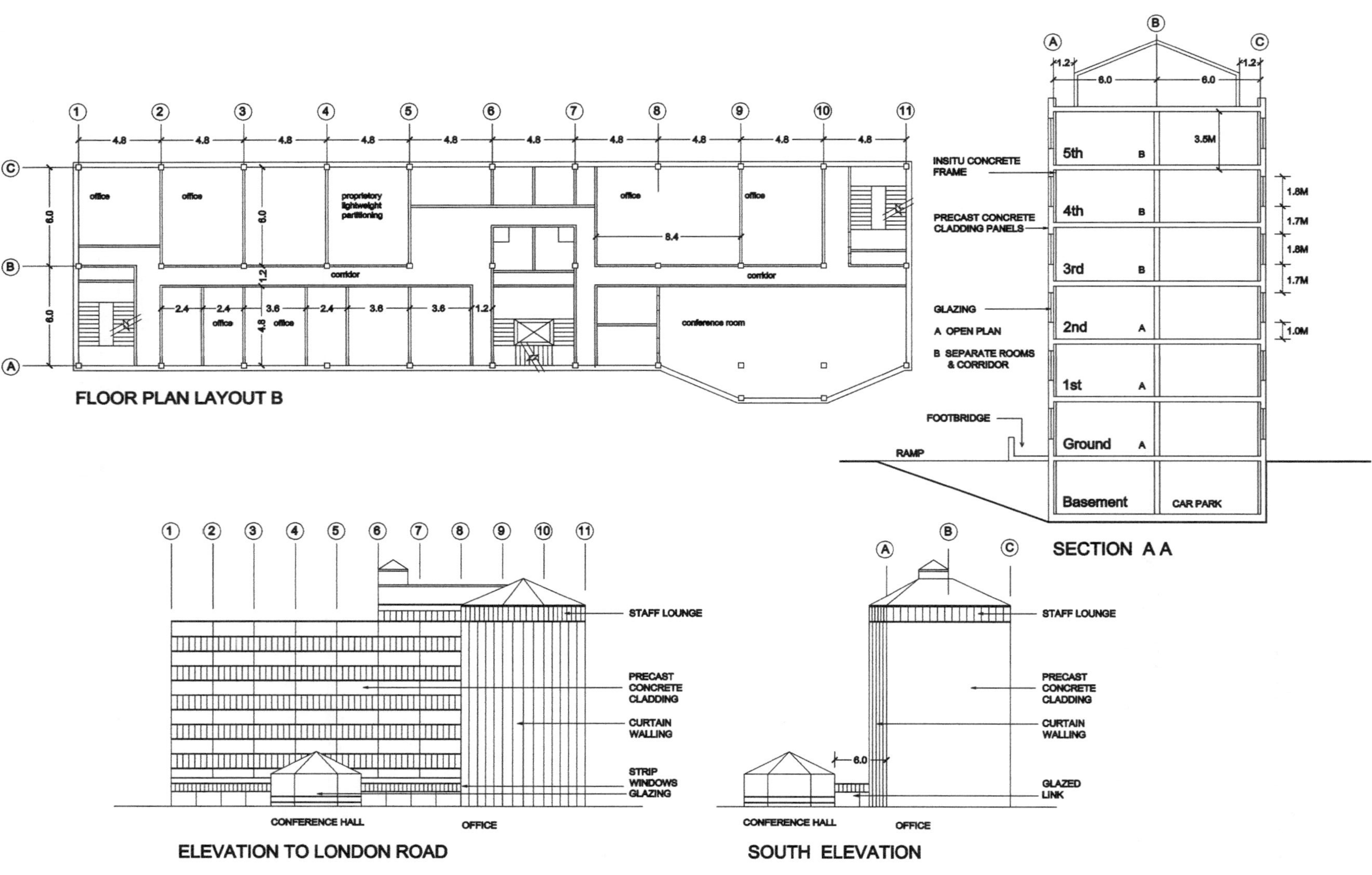

Reproduced by permission of The College of Estate Management.

References

Ashworth, A. (2010) *Cost Studies of Buildings*, 5th edn. Harlow, UK: Pearson Prentice Hall.

Ashworth, A. and Hogg, K. I. (2007) *Willis's Practice and Procedure for the Quantity Surveyor*, 12th edn. Oxford, UK: Blackwell.

Atton, W. (1975) *Estimating Applied to Building*, 3rd edn. London: George Godwin Ltd.

BCIS (2012) *Elemental Standard Form of Cost Analysis*, 4th edn. London: RICS Publications.

Building and Allied Trades Joint Industrial Tribunal (2010) Working Rule Agreement. London: Building and Allied Trades Joint Industrial Tribunal.

Davis Langdon (2011) *Spon's Architects' and Builders' Price Book 2012*, 137th edn. London: Spon Press.

Gagne, R. (ed) (2002) *The Conditions of Learning*. Austin, TX: Holt, Rinehart & Winston.

Hackney, J. W. (1992) *Control & Management of Capital Projects*, 2nd edn. New York: McGraw-Hill.

RICS (2007a) *The RICS Code of Measuring Practice: 6th edn*. London: RICS Publishing.

RICS (2007b) *The RICS New Rules of Measurement NRM 2: Detailed Measurement for Building Works*. London: RICS Publishing.

RICS (2007c) *RICS 'Definition of Prime Cost of Daywork Carried Out Under A Building Contract'* 3rd edn. London: RICS Publishing.

RICS (2011) *Quarterly Review of Building Prices*. London: RICS Publishing

RICS (2012) *The RICS New Rules of Measurement NRM 1: Order of Cost Estimating and Cost Planning for Capital Works*, 2nd edn. London: RICS Publishing.

RICS Davis Langdon (2010) *A Survey of Building Contracts in Use*, 11th edn. London: RICS Publishing.

Seeley, I. H. (1976) *Building Economics*, 2nd edn. London: Macmillan.

Stratton, C. (2003) *The Plant Hire Investment Report Published 1993–2003*. Corsham, UK: PHIR.

Wood, D. (2001) 'Scaffolding, contingent tutoring and computer supported learning', *International Journal of Artificial Intelligence in Education*, 12: 280–92.

Working Rule Agreement, Construction Industry Joint Council, London, UK, 2010.

Estimating and Cost Planning Using the New Rules of Measurement, First Edition. Sean D.C. Ostrowski.

Index

Estimating and Cost Planning Using the New Rules of Measurement, First Edition. Sean D.C. Ostrowski.

Printed and bound by CPI Group (UK) Ltd, Croydon, CR0 4YY
09/06/2025
14686006-0002